Analysing Environmental Data

Allan Pentecost

Longman

Pearson Education Limited
Edinburgh Gate, Harlow
Essex CM20 2JE
England
and Associated Companies throughout the World

© Pearson Education Limited 1999

First published 1999

ISBN 0 582 31058 X

British Library Cataloguing-in-Publication Data
A catalogue record for this book is
available from the British Library.

Library of Congress Cataloging-in-Publication Data
A catalog entry for this title is
available from the Library of Congress.

Set by 35 in $9\frac{1}{2}$/12pt Concorde BE
Produced by Addison Wesley Longman Singapore (Pte) Ltd.
Printed in Singapore

Contents

13 An introduction to modelling   136

14 An introduction to toxicity testing   144

15 An introduction to multivariate analysis   152

16 Questionnaires   167

17 Sampling and experimental design revisited   176

Preface

This book is intended as a primer to more advanced courses and has been written specifically for those interested in environmental problems and taking a first, or refresher course in data analysis at university. Some sections of the book will be familiar to those of you already possessing statistics texts. Chapters 3 to 12 contain a straightforward account of the normal distribution, frequency analysis, two-sample tests, analysis of variance and correlation/regression techniques. Example calculations are provided throughout using data collected from a wide range of environmental studies.

Although many teachers feared that the advent of mouse-driven software packages would prevent their users from gaining an understanding of the principles behind statistical methods, this has been unfounded. On the contrary, packages have removed much of the drudgery of repetitive calculation in data analysis while stimulating interest in the data themselves and releasing time for students to design better experiments and delve more deeply into numerical methods. Needless to say, my own students have also benefited from a course where at least a few examples of each method are worked through with a simple pocket calculator. This has been found to improve numeracy skills, which often need honing after a break from school and help to develop experience and confidence in handling numerical data.

The presentation style should help to clarify the data analysis examples in the main text. For instance, the critical tables and critical values drawn from them are shown thoughout in **bold** type to distinguish them from statistics calculated from the data. With the increasing use of computer packages, it was felt important to include additional examples using *Minitab 12.0*, the most recent version of this package at the time of going to the press. These analyses are fully explained with printouts from the package. The company that created this package can be contacted at the following address:

Minitab Inc.
3081 Enterprise Drive
State College, PA 16801 USA
tel: 814 238 3280
fax: 814 238 4383
e-mail: Info@minitab.com
URL: http://www.minitab.com

Finally I am grateful to all of those individuals who have assisted in the production of the text. These include the two independent reviewers who made many useful suggestions and gave advice to help improve the text. I am also grateful to Drs. Nyholm and Welch for permission to make use of their original data and particularly to my wife who prepared the drawings to illustrate some of the examples. Most of the exercise examples have been tested by several groups of Environmental Science students at King's College London and it is to be hoped that they are free of error.

Allan Pentecost
17.12.1998

CHAPTER 1

Introduction

This text is designed for undergraduate environmental science courses on data analysis and statistics. The need to understand, analyse and interpret numerical data is paramount in all scientific disciplines. The advent of the desk-top computer has brought with it powerful analytical techniques which were difficult to access even a decade ago. Increasing use is made of these methods in environmental science and with their widespread availability there is the attendant need for teachers to continuously update courses to include them.

A glance through environmental science articles published in the early 1960s will reveal comparatively few papers using statistical analysis. Since that time, surveys indicate that the number of publications using statistical methods has more than doubled, and often include advanced multivariate methods which were unheard of 40 years ago. There is nothing to indicate that this increasing trend is diminishing and the environmental scientist needs to be aware of the advances in data analysis and the significance of these advances in environmental studies. Therefore it is important that the portfolio of the modern environmentalist contains the essentials of hypothesis testing and data analysis.

Many environmental problems are concerned with changes in time. For example, we may need to find the answer to 'how will a plant community respond to a pollution event over the coming years?'. Such questions often need to be addressed as part of an environmental impact assessment and can only be properly answered using a logical experimental framework which includes a *null hypothesis*, a statement which, in this case, implies invariance with time. It is then up to the researcher, using a range of experimental and statistical tools, to test the hypothesis and make a reasoned assessment.

In addition to the basic analytical principles and methods, which are essential to analyse numerical data, some more specialised chapters are included of particular relevance to environmental science. These include a section on the design and analysis of questionnaires which in our own experience at King's College are often requested by final year project students. In Chapter 13 an introduction to modelling is given, which in many respects is a natural extension of regression analysis and some simple examples of environmental models are provided. Also included is an introduction to toxicity testing and analysis.

Undergraduate primers on data analysis do not normally include material on multivariate methods but the author considers the subject so important that a brief overview of the major techniques has been included. Indeed the author feels that this topic ought to be included in all undergraduate environmental science degree programmes. There is great demand for people qualified in the use of multivariate methods and the potential for applying these methods still appears to be largely unappreciated in the environmental field, while it has expanded enormously in the social and biological sciences. Most environmental measurements are taken outside the confines of a laboratory where climatic variables (e.g. windspeed, temperature, irradiance) change continously. Multivariate methods are particularly appropriate in these situations, whereas in the laboratory where there is deliberate control of variables, uni- or bivariate analyses are usually more suitable.

To make good use of the book, a basic level of mathematics is assumed, such as that provided in the UK by the GCSE examination. There is no substitute for a command of basic arithmetic, but the material presented here needs no more than a rudimentary knowledge of algebra. However, the ability to calculate logarithms and antilogarithms will be a distinct advantage since these are used widely in data analysis. To fully appreciate most multivariate methods, an understanding of matrix algebra is needed, but this is not essential, and little direct reference is made to matrix algebra in the text.

CHAPTER TWO

Conducting and designing experiments

Most scientific progress has been achieved through experimentation. The first experiments must have been undertaken through chance observations, such as the production of fire by frictional heating. Following the observation, experiments would be made to discover an efficient way of making and retaining the fire. Even today chance obervations can lead to interesting discoveries through experimentation, such as the recent thunderstorm 'sprites'. However, most experiments are undertaken nowadays to test a hypothesis for which there is a history of observation.

A *hypothesis* is a statement relating to an observation which may be true but for which a proof (or disproof) has not been found. Thus an experiment may be considered as an operation which could ultimately lead to the proof or disproof of a hypothesis.

Individual experiments rarely lead to the proof of a statement, but if they do, then the statement or 'law' is often regarded as a *theory*. The terms hypothesis and theory are distinct, but they are frequently muddled and confused even within scientific circles. This is partly because people involved with the 'hard' sciences (mathematics, physics) often develop theories from premises or other theories which do not require experimentation.

Undertaking experiments invariably leads to measuring operations required to test the hypothesis in question. Most experiments are planned, or designed, either according to a previously used protocol or by undertaking a small 'pilot' experiment.

Let us look at an experiment which might be undertaken by an environmental scientist. The researcher wishes to find out whether the uptake of lead (Pb) in solution by a crustacean *Asellus* is influenced by the amount of calcium in the water. The experiment is being undertaken because measurements taken in streams seem to indicate that in 'hard water', i.e. a water rich in calcium and magnesium, the amount of lead taken up by the crustacean is reduced. It has been hypothesised that, in waters containing

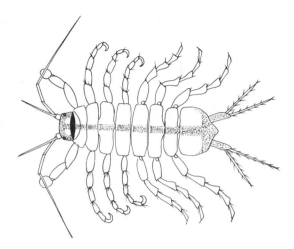

elevated calcium, calcium competes with lead for adsorption sites in the crustacean, therefore reducing its uptake.

Clearly the material required includes some *Asellus*, some water containing lead and calcium, and some equipment for measuring lead in the *Asellus*. Accordingly, 40 *Asellus* were collected from a short stretch of river and maintained in a small aquarium. One method to test the hypothesis might be to place all 40 crustaceans in a solution containing both lead and calcium and analyse them for their lead content after a specified time. We could then turn to the literature and compare our results with those of previous studies. It might prove to be an interesting experiment, but it is not a satisfactory way of proceeding. Two things are wrong with the experiment. First, we have no idea how much lead was present in the *Asellus* at the beginning of the experiment, and thus no idea how the lead content might have changed after the procedure. Second, although there may be literature we can consult, it is unlikely that the earlier experiments were conducted under the same conditions as ours (i.e. temperature, time of incubation, composition of medium), making any comparisons meaningless. By consulting the literature we may gain useful tips on how to conduct the experiment, but with wild animal samples we can never reproduce the conditions precisely. To overcome the problem, we need a *control* experiment with which to compare our results.

A better procedure would be to divide the *Asellus* into two or more groups, use one group as a control containing no calcium and incubate the others with calcium-containing solutions. Here another problem is encountered, as it will be obvious that the individual animals will differ in age, size and also probably in sex. A natural **sample** is most unlikely to consist of genetically identical animals so that each *Asellus* will probably take up lead at different rates. This means that the rate of uptake of lead is likely to be different between the individuals. This intrinsic variability is the result of the non-uniform nature of the biological world (and the world in general). Other experiments could be conducted to estimate the intrinsic variability but they would be time-consuming and costly. Instead, the differences between individuals could be smoothed out by picking out individuals at random and assigning them to one of the two or more treatments. Each individual has an equal chance of being selected for any treatment and, providing we have a sufficient number of individuals to choose from, the variability in each of the treatments should be approximately the same.

With 40 *Asellus* the experiment can be planned in several ways but it is important to organise the study to best suit the hypothesis under test. We could divide the sample in two with 20 randomly selected *Asellus* in the Ca– (control) group and 20 in the Ca+ group with a fixed concentration of lead. Or, we could take four sets of ten animals and have one control group and three sets exposed to different levels of calcium. Without some prior knowledge of the variability of lead uptake by *Asellus* it is not possible to choose the best option. Where we are in a state of ignorance it would be sensible to use as many replicates as possible and treat the first experiment as a 'pilot' study, assigning 20 *Asellus* to the Ca– and 20 to the Ca+ groups. Although this is sound practice, the use of large numbers of replicates can sometimes lead to further problems, increasing the variability and defeating the object of the exercise. If for example we decided to place each *Asellus* in the beaker of water, we would need a considerable amount of space to accommodate all 40 beakers. The area assigned will probably be subject to small variations in temperature and light intensity, which could influence the animal's behaviour. Setting out the beakers also becomes a part of the design and ideally should be done as a *randomised block* (Chapter 17). Also, it would take a considerable time to set up all of the beakers, and as this is a timed experiment it would be difficult to have precise timing for each of the 40 incubations. Clearly the experimental space is itself subject to its own variability, which must be taken into account. One way of reducing the variability would be to place the two samples of *Asellus* in two large containers with well mixed solutions. An important principle of experimental design, where time and space is limited, is to keep the number of replicates and manipulations to a manageable size.

After the *Asellus* have been incubated with the solutions they must be rapidly killed and removed from the solutions to prevent further lead uptake. In an experiment of this nature there may be insufficient time to complete the study in one day and the animals will need to be preserved in a chemical or by freezing to prevent decay and possible release of lead. If the animals are simply left overnight in a warm laboratory, decay will inevitably set in without preservation; although it could be argued that if the control and Ca+ groups are left under the same conditions the effect will even out. If we were simply looking for differences in lead uptake and were not interested in the absolute amount of lead in the two samples this may be acceptable, but there is still a potential problem – those incubated with calcium may decay at a different rate and it would be unwise to assume otherwise without conducting a separate experiment.

The next stage of the study would involve converting the *Asellus* lead into a soluble form so that it can be measured. This is normally achieved by dissolving the animals in a hot, strong acid mix. With 40 animals this is time-consuming since the operation has to be conducted carefully in a fume cupboard. The final solution will be made up to a fixed volume and may be analysed using atomic absorption spectrophotometry, which involves spraying a small volume of the solution (about 0.1 ml) into a hot flame so that the lead atoms are released from chemical combination. The atoms absorb light at particular wavelengths and the machine can be tuned to a particular element. The instrument is incapable of making absolute measurements and so it has to be calibrated by spraying a series of standard lead solutions of known concentration. Lead in the unknown solutions is then determined from a calibration graph and the final results can be expressed as microgrammes of lead per animal per hour, allowing comparison of the two groups.

The analytical stage of the experiment also involves measurements, which add a further element of variability to the final result. In atomic absorption spectrophotometry there is a significant risk of contaminating the samples with the concentrated lead standard used to calibrate the equipment. Also, because of small differences in flame characteristics and electronic noise the instrument itself will not provide precisely the same result if the same individual sample is run through several times. In addition, sample preparation is prone to several sources of error. These analytical uncertainties present themselves in the results as extrinsic variability. Both intrinsic and extrinsic variability can be controlled to some extent by the experimenter and the magnitude of both should be kept as low as is practicable.

Prior to performing the experiment it is important to consider how the hypothesis is presented. If the results show that the presence of calcium does reduce the uptake of lead in *Asellus* then we might be tempted to state that the hypothesis that calcium inhibits the rate of uptake of lead is proved. This may indeed be the case, but to prove a hypothesis on the basis of a single experiment is unwise. If proof is admitted then, in all other situations, lead uptake by *Asellus* will be inhibited by calcium. Our experiment paid no attention to the particular calcium concentration used and it is quite possible that at some calcium concentrations lead uptake is not affected. To circumvent this problem, which is common to all experimental procedures, it is customary to propose a *null hypothesis*. This hypothesis is stated in terms of a disproof so that no assumptions are then required to cover all possible cases. In our case the null hypothesis would be that 'the presence of calcium has no effect on the uptake of lead by *Asellus*'. Then, if an inhibitory effect was found we could state that the experiment did not support the null hypothesis, leading to the acceptance of the alternative explanation. Lead uptake may well be reduced in the presence of calcium and further experiments should be undertaken. It is of interest to consider the procedure if the two samples indicated the same rate of uptake. In this case there is clearly no evidence of calcium inhibition and the null hypothesis would be accepted.

The original observation was based on evidence from studies of 'hard water' and these waters often contain high levels of magnesium and frequently possess a high pH. Therefore the original observations may also be consistent with hypotheses relating to these additional factors, leading to further experimentation.

In the physical and chemical sciences the results of an experiment may be so precise that they may be clearly interpreted from a graph or table. In many other branches of science, including environmental science, the variation is frequently large enough to obscure any obvious differences between treatments and statistical analysis is required. These analyses are undertaken with reference to probabilities, and the null hypothesis is more formally stated in terms of probabilities. This topic is taken up in Chapter 5. In the *Asellus* example, a comparison between the Ca+ and Ca− groups could be made using the ordinary *t* test (Chapter 8). If, however, it was possible to divide the sample in two and 'pair off' each *Asellus* with another of the same age, size and sex, a paired *t* test would be more reliable. If the sample was randomly split into three or more subsamples of similar size, which were treated with different levels of lead, then a single classification analysis of variance (ANOVA) and regression could be applied (Chapters 10 and 12). It is therefore important to understand which statistical methods are applicable to the experimental design. Ideally this should be done when the experiment is planned.

Key notes

- Before beginning an experiment, think carefully about what you wish to demonstrate and present the problem as a hypothesis to be tested.

- Review the experimental procedures used by others so that the less obvious pitfalls can be recognised.

- If possible, conduct a pilot experiment to iron out difficult procedures and obtain information to improve the precision of the final results.

- Ensure that the size of the experiment is manageable in terms of cost, time and the space needed for the manipulations.

- Make sure that the experiment is designed within a statistical testing framework.

CHAPTER THREE

Descriptive statistics

3.1 Measurement theory

We all make measurements throughout our lives, in order to compare the relative size of objects we see, hear or touch. As scientists we use measurements routinely in laboratory or field work, and by assigning numbers to measurements we can record data for future use, and perform useful operations such as addition, subtraction and the calculation of averages. The advantages of these operations were appreciated by the early farmers and builders who were the founders of our civilisation. For example, by measuring the length of the sides of a squared block of stone its weight can be established and its mode of transport and value determined. New information can therefore be obtained using a simple mathematical operation such as volume estimation.

While many everyday objects can be measured with a ruler or similar device, it soon became apparent that many objects of interest could not be readily measured. In the early years of astronomy, stars were found to differ in their brightness but there was no way of accurately comparing them. They could be placed on a relative scale, but an absolute value could not be assigned. It became apparent that measurements could be taken with different degrees of veracity, yet still provide valuable information. Theories of measurement were founded, resulting in the construction of a scale of *levels of measurement*. Four basic levels of measurement were soon recognised and are described below.

3.1.1 The nominal level of measurement

The **nominal** level of measurement is sometimes termed the 'classificatory' level: numbers or other symbols are used to classify an object or a character into a range of categories. Consider a questionnaire with 100 respondents, 50 male and 50 female, who were asked whether they agreed or disagreed with the statement 'All domestic waste should be recycled'. The replies are presented below in two rows and two columns.

	Agree	Disagree
Males	8	42
Females	13	37

The results could have been presented as a bar chart (Chapter 4), but the table, known as a **cross-tabulation**, facilitates statistical analysis. The results of the survey show that the responses of males and females were similar and the majority of people disagreed with the statement. These measurements are made at the nominal level: they are simply 'counts' of a recorded characteristic. The order in which the counts are given has no significance and as counts they are given as whole numbers.

The nominal level is the weakest level of measurement because the counts (or **frequencies**) obtained convey limited information. The measurements relate only to how often a particular character appears. When we produce a series of counts for each character, the descriptive terms we apply to the set of frequencies are the most frequent class (usually referred to as the **mode**) and the number of categories. Statistical methods for analysing such data include the χ^2 and G tests (Chapter 6).

3.1.2 The ordinal level of measurement

It is often found that objects of interest do not differ merely in shape or type, but vary in a way that allows them to be placed in some form of order. For example, some light-sensitive paints are available which change colour as the ultraviolet radiation dose absorbed by them increases (e.g. Jones, 1982). Painted discs can therefore be fixed to surfaces and after a certain time has elapsed may be collected and the colours compared. The discs can be placed in order from darkest to lightest, with the lightest having received the highest dose of radiation. Thus, the results can be presented in an orderly fashion with the lightest given the lowest score and the darkest the highest score. This level of measurement provides more information than the nominal scale because the measurements can be **ranked**. More familiar examples are the ranks in an army, where commands are handed down from one rank to the next, e.g. sergeant–corporal–private. They have also been used to compare air pollution damage to tree foliage. Ordinal measurements are familiar to social scientists who often need to consider characters whose value cannot be directly assessed, such as socioeconomic status (working class, middle class, aristocracy, etc.). Descriptive statistics at the ordinal level include again the mode (the most frequent class), the **median** (the middle value, used as a measure of 'central tendency') and the **interquartile** range (a measure of 'spread' or **dispersion**). These terms will be described later.

3.1.3 The interval and ratio levels of measurement

The **interval** level of measurement is the most frequently used today: a scale is involved and the distance between any two numbers on the scale is a known or measurable quantity. Familiar examples of interval measurements are time as measured by a clock, distance in metres or mass in kilogrammes. On an ordinal scale, the difference between two measurements is not a known quantity because there are no accurate criteria with which to measure it, or a measuring device may not be available to make an accurate measurement (as in the paint example). Thus the interval scale is characterised by a common and constant unit of measurement. The ratio of any two intervals is also independent of the measurement unit applied. This means that temperature can be recorded on different interval scales (e.g. Celsius, Fahrenheit) and be readily interconverted.

Statisticians recognise a special form of interval scale known as the **ratio** scale. This scale differs only in the possession of an absolute zero value. Distance and mass are measured on this scale, which means that the measures can be subjected easily to arithmetic procedures. For example, I weighed 68.72 kg before Christmas last year and 69.86 kg immediately afterwards. My weight therefore increased by $[(69.86 - 68.72)/68.72] \times 100\%$ or 1.66%. An increase in temperature on the Celsius or Fahrenheit scales could not be subjected to the same procedure because they are not absolute scales. However, if the temperatures had been measured on the Kelvin scale, they could. For both levels the important descriptive statistics are the **mean** (for a measure of **location**) and **standard deviation** (for a measure of spread or dispersion).

Modern measuring devices often allow measurement of mass to be made to several decimal places. For example, it is possible to weigh an insect to at least five decimal places with a torsion balance (e.g. as 0.200 01 mg). Such a device is termed a **precision** instrument. However, if the instrument has not been calibrated, the result will not necessarily be **accurate**, i.e. provide a true value of the insect's mass. It should also be noted that precision measurements do not always involve numbers containing many decimal places. If I count the number of lorries passing down a lane during a specified time period (e.g. 1 hour) and find a total of 15 the value should be both precise and accurate.

This leads on to one final observation – the data we collect can be of two forms, consisting of whole numbers only or of numbers on a continuous scale, including fractions or decimals. Whole number scales normally consist of counts of objects, attributes and characters and are referred to as **discontinuous measurements**, while measurements which can in theory take any value on a continuous scale are called **continuous measurements**. It is rarely, if ever, possible to measure continuous variables with great precision and the last digit of the value implies the precision achieved. For example, a mass of 14.7 g implies that the true mass lies between 14.65 and 14.75 g. In scientific work the highest degree of precision is sought for measurement, and often it is necessary to take measurements over a wide range of scales. This has resulted in the evolution of special terms for decimal multiples of

Table 3.1 Decimal multiples of length and mass

Scale	Length	Mass
10^{-9} nano	nanometre, nm	nanogramme, ng
10^{-6} micro	micrometre, µm (micron)	microgramme, µg
10^{-3} milli	millimetre, mm	milligramme, mg
10^{-2} centi	centimetre, cm	centigramme
1	metre	gramme
10^{3} kilo	kilometre, km	kilogramme, kg
10^{6} mega	–	tonne

scale, as indicated in Table 3.1 for length and mass. Variables such as length and mass are sometimes referred to as having *extensive* properties, in contrast with variables such as temperature and pressure which have *intensive* properties.

3.2 Measurements of location

There is a need with most measurements to provide a short summary of the information they convey and we have already seen that the measurements themselves, through their levels, provide at least some information. However, it is also important to have a single number, based upon the measurements, which provides a typical value for the sample, or a measure of centrality. There are two measures of central tendency, or location: the *mean* and the *median*.

3.2.1 The mean

The mean is the average value of a data set and is obtained by adding the individual values and dividing by the number added (n). Thus

$$\text{mean} = \text{sum of values}/n$$

BOX 3.1: EXAMPLE

Measurements of wind speed (km h^{-1}) were obtained at a meteorological station at 1 h intervals for 8 h. Find the mean wind speed from these data:

$$38 \quad 24 \quad 12 \quad 17 \quad 28 \quad 32 \quad 19 \quad 16$$

The number of measurements $n = 8$, giving

$$\text{mean} = (38 + 24 + 12 + 17 + 28 + 32 + 19 + 16)/8$$

$$= 186/8 = 23.25 \text{ km h}^{-1}$$

BOX 3.2: EXAMPLE

During the same period ($n = 8$), air temperature (°C) was recorded at the meteorological station as

$$3.4 \quad 2.4 \quad 1.8 \quad 0.4 \quad 0.0 \quad 0.7 \quad -1.2 \quad -3.2$$

$$\bar{x} = (3.4 + 2.4 + 1.8 + 0.4 + 0.0 + 0.7 - 1.2 - 3.2)/8$$

$$= 4.3/8 = 0.5375 \; °C$$

Note that the zero value must be included in the calculation as it represents one of the **datum** points.

In statistics, the sample mean is given its own special symbol, $\bar{x}$, which is used more or less universally. However, the Greek letter μ is used for the *population mean*, and will be considered later.

Calculating the mean presents no difficulties, providing all the numbers are positive. If some negative numbers are included, more care needs to be taken as these will be subtracted rather than added in the summation. This may occur if temperatures around freezing are measured in degrees Celsius, as shown in Box 3.2.

3.2.2 Sigma notation

The Greek letter S is written as Σ and is used widely in data analysis. If an individual measurement from a sample is defined as x, then Σx means 'sum of the values of x'. Thus, for the four relative humidity values below

$$\Sigma x = 96 + 91 + 95 + 90 = 372$$

where x represents the values 96, 91, 95 and 90. The mean can then be written concisely as

$$\bar{x} = \Sigma x/n$$

Occasionally, individual values of a data set need to be distinguished. If the humidities were recorded in the order 96, 91, 95, 90, each value may be given its own symbol by using a **subscript**: $x_1 = 96$, $x_2 = 91$, $x_3 = 95$ and so on to x_i where i represents the ith value in the set

3.2.3 The median

The median represents the 'middle value' of a data set. To find the median the data must first be ranked from smallest to largest, and the mid-point is the median. As an example, consider nine measurements of soil temperature (°C)

$$8.2 \quad 8.7 \quad 9.5 \quad 7.2 \quad 8.0 \quad 8.9 \quad 7.6 \quad 6.9 \quad 9.0$$

To find the median, the measurements are first rearranged from lowest to highest, a technique known as ranking. The numbers in italics below are the *order statistics*, which show us the rank position of each measurement.

6.9	7.2	7.6	8.0	8.2	8.7	8.9	9.0	9.5
1	*2*	*3*	*4*	*5*	*6*	*7*	*8*	*9*

Since the median is the middle value, we take the datum point represented by order statistic 5. Since there are four values above and four below, order statistic 5 represents the middle value. Therefore the median is 8.2.

Calculation of the median is a simple procedure if the data string is not too long and the number of measurements, n, is odd, as above. If the number of measurements had been even, then the median would lie between two of the values. In the case of the four relative humidities ranked below, the median will lie between 91 and 95.

90	91	95	96
1	*2*	*3*	*4*

In such cases, the median is taken as the mean of the values lying either side. In this case, the median is $(91 + 95)/2 = 93$. With larger data sets, the location of the median (i.e. its order statistic) can be found using the formula

$$\tilde{x} \text{ (order statistic)} = (n + 1)/2$$

Remember that the order statistic gives only the location of the median, and not its value. The formula works whether n is odd or even. Note the symbol used for the median ($\tilde{x}$, read as 'x tilde').

The median is often used with the mean to provide two measures of location for a single sample. In publications, the mean is quoted much more frequently than the median, but for data whose distribution is **skewed**, or with outliers, the median is a better measure of central tendency. The median is also the only measure of location that can be used with ordinal data, whereas with interval/ratio data both measures of location can be used.

3.3 Measurements of dispersion

A measurement of location tells us something about the position of our data in numerical terms, which permits us to make a rough comparison with other data. On its own it tells us nothing about the range of the values within the sample. The range or 'spread' of the measurements may be very important. For example, a paper mill may not be allowed to discharge an effluent containing in excess of 50 ppm suspended solids. If the average value discharged from the works has been found to be 20 ppm, we cannot deduce that the effluent always complies with the law. For this, a knowledge of the range of values must be obtained. As with measurements of location, there are several measures of 'spread' or dispersion.

The simplest measure of dispersion is the **range**, which is the difference between the highest and lowest value within a sample. With a series of suspended solids determinations (ppm) 15, 45, 11, 22, 28, 29, 31, 41, 33, the highest and lowest values are 45 and 11, giving a range of $45 - 11 = 34$. The highest and lowest values themselves are also often quoted in publications, where the range itself is implicit.

The range cannot reveal structure within the body of measurements and this too is often important. For instance the suspended solids measurements may show that the 50 ppm limit is exceeded only once for one sample while another sample might show it to be exceeded several times. Thus the spread of measurements is not well described by the range.

3.3.1 The standard deviation

The standard deviation provides the best measure of dispersion for interval/ratio measurements and is perhaps the most important of all statistical measurements after the mean. Its use arose out of the need to provide an averaged measurement of **deviation** from the mean. In a sample of measurements, each individual deviation from the mean is written as

$$x - \bar{x}$$

Look at the five rainfall measurements, whose mean is 7, tabulated below.

Rainfall (mm) x	Deviation from the mean $x - \bar{x}$	Squared deviation of the mean $(x - \bar{x})^2$
12	$12 - 7 = 5$	$5 \times 5 = 25$
0	$0 - 7 = -7$	$-7 \times -7 = 49$
2	$2 - 7 = -5$	$-5 \times -5 = 25$
5	$5 - 7 = -2$	$-2 \times -2 = 4$
16	$16 - 7 = 9$	$9 \times 9 = 81$

A good measure of dispersion might be considered to be the average deviation from the mean, which is obtained by adding all of the deviations and dividing by $n = 5$. If this is attempted you find that the sum $5 - 7 - 5 - 2 + 9 = 0$, which is unhelpful. The positive and negative deviations cancel so the method cannot be used. An alternative is to make all the deviations positive and then average. This was once widely practised, but there are good theoretical reasons for making the deviations positive by squaring them instead.

The **sum of the squared deviations** is then $5^2 + (-7)^2 + (-5)^2 + (-2)^2 + 9^2$ or $25 + 49 + 25 + 4 + 81 = 184$. The mean of this sum is $184/5 = 36.8$.

Again, sigma (Σ) notation can be used to simplify the process. With this the sum of the squared deviations becomes

$$\Sigma(x - \bar{x})^2$$

and the mean sum of the squared deviations is written

$$\Sigma(x - \bar{x})^2/n$$

This quantity might be considered to be an adequate measure of dispersion. It could be used as such, and it is an important statistical quantity called the **population variance**. However, the effect of squaring the deviations makes them much larger than the original unsquared deviations and to remove this effect they are unsquared by taking the *square root* of the sum of the squared deviations. This is the standard deviation.

$$\text{population standard deviation } (s) = \sqrt{[\Sigma(x - \bar{x})^2]/n}$$

3.4 Two important modifications

As stated above the standard deviation is a very important measure but it is not normally calculated in the above manner. Two modifications are applied. The first provides a better estimate of the population standard deviation and the second provides a more efficient method of calculation.

1. The mean squared deviation was obtained by dividing the sum of the squared deviations by n, the number of measurements. However, mathematicians have found that by dividing the sum by $n - 1$ (also known as the degrees of freedom, see Chapter 6) rather than n gives a better (unbiased) estimate of the population standard deviation. This becomes more important when we consider inferential statistics later on. The effect of dividing the sum by $n - 1$ rather than n is to make the standard deviation slightly larger. However, as n increases the difference declines rapidly in importance.
2. The method shown in Box 3.3 should always be used to calculate the standard deviation if an electronic calculator 'sd' function is unavailable. It is done in three steps as illustrated.

BOX 3.3: CALCULATION OF THE STANDARD DEVIATION

First, obtain the sum of the squared deviations using the formula

$$\Sigma x^2 - \frac{(\Sigma x)^2}{n}$$

Be careful here, you may be misled into thinking that the mean deviation is being calculated because the right-hand term is divided by n. This is not the case: the mean is calculated in the second step.

For the humidity data above

$$\Sigma x^2 = 12^2 + 0^2 + 2^2 + 5^2 + 16^2$$

$$= 144 + 0 + 4 + 25 + 256 = 429$$

$$\Sigma x = 12 + 0 + 2 + 5 + 16 = 35$$

$$(\Sigma x)^2 = 35^2 = 1225$$

Thus the sum of the squared deviations is

$$429 - 1225/5 = 184$$

Next, obtain the mean sum of the squared deviations (the variance) by dividing by $n - 1$ (degrees of freedom). Note that all sums of squares are positive numbers. If the result is negative then a mistake has been made.

$$\frac{\sum x^2 - \left(\sum x\right)^2/n}{n - 1}$$

For this example, the mean sum is

$$184/(n - 1) = 184/4 = 46.0$$

Finally take the square root to yield the standard deviation

$$\sqrt{(46.0)} = 6.782$$

One of the commonest mistakes in statistical calculations is to obtain the sum of the squared deviations but then forget to divide the sum by $n - 1$ before obtaining the standard deviation.

3.5 Percentiles, quartiles and the interquartile range

The median has already been shown to represent the 'middle' or 'half' value of a set of data taken at ordinal or interval level. By the same token, a data set can be divided equally into four, resulting in three markers, a *lower quartile*, the median and an *upper quartile*.

The method used to find the positions of the **quartiles** is the same as that used for the median and the difference between the upper and lower quartiles is called the *interquartile range*. The lower and upper quartiles are written Q_1 and Q_3, respectively, and the interquartile range (IQR) is $Q_3 - Q_1$. Clearly, the interquartile range contains half of the measurements taken and is centred upon the median and thus a good measure of dispersion. It does not possess the special mathematical properties of the standard deviation, but it is more resistant to the presence of outliers in the data.

3.5.1 Percentiles

It is often useful to determine, within a set of data, where a particular value lies when the data are ranked from lowest to highest. For example, we might be informed for a particular weather station that a temperature of 18.1 °C was the 95th percentile. This means that, when all the temperature measurements were ranked, 95% of them had a temperature between the lowest recorded and 18.1 °C. The mth percentile represents a point which separates the lower m% of the measurements from the upper $(100 - m)$%. The median represents the 50th percentile and the upper quartile the 75th percentile. The method used to find the median and quartiles can also be used to calculate percentiles. Quartiles and percentiles are sometimes referred to as *quantiles*. Another term, *decile*, is also occasionally used to describe a data set split up into tenths rather than quarters.

Descriptive statistics with Minitab

With the following data entered into column 1 (C1) of a worksheet

12, 23, 56, 32, 13, 45, 8, 14, 17, 32, 21, 53, 21, 26, 29

the operation from the menu bar

Stat > Basic Statistics → Display Descriptive Statistics

will provide a *dialog box* in which C1 is entered in the 'Variables' box, followed by clicking 'OK'. The output is shown below and provides the number of measurements, *n*, the mean, the median, the 'trimmed mean' (Tr Mean), which is obtained by averaging the data between the fifth and 95th percentiles only, to remove the effect of outliers when they are present, the standard deviation, the standard error (see Chapter 9), the minimum value, the maximum value and the lower (Q_1) and upper (Q_3) quartiles, permitting calculation of the interquartile range.

Variable	N	Mean	Median	TrMean	StDev	SE Mean
C1	15	26.80	23.00	26.00	14.72	3.80
Variable	Minimum	Maximum	Q1	Q3		
C1	8.00	56.00	14.00	32.00		

Key notes

- Measurements are taken at four levels, nominal, ordinal, interval and ratio.

- Nominal level measurements classify the characteristics of a sample.

- Ordinal level measurements allow observations to be ranked in order from lowest to highest.

- Interval level measurements are placed on a scale of values, where the scale represents an accurately measurable quantity.

- Ratio level measurements are interval level measurements with a true zero point, e.g. the Kelvin temperature scale.

- Measures of location provide a central value for observations and include the median, $\tilde{x}$ and mean, $\bar{x}$.

- Measures of dispersion describe the range of observations and include the absolute range, the interquartile range and the standard deviation(s).

- The mean and the standard deviation are the most important descriptive statistics.

CHAPTER FOUR

Presentation of data

4.1 Tables

Data can be presented as tables or graphs. Tabulated data provide the most accessible and transferable form for measurements, and with descriptive statistics (Chapter 3) provide a useful summary of experimental results. With larger data sets tables become unwieldy and in any case usually fail to reveal the structure of the data. Nevertheless, in research papers, you will find that data are presented in tabulated form at least as often as in graphical form. Tabulated data must be self-explanatory with sufficient spaces between the numbers to avoid confusion. Columns and rows must be labelled and the units in which the measurements are made must be specified (preferably SI units, Système International d'Unités). Fundamental SI units are the kilogramme, metre, second, kelvin (temperature) and mole. The legend should be fully explanatory and the number of significant figures provided in the table must not exceed the accuracy of the method used to take the measurements. An example is shown in Table 4.1.

The altitude values in the table could also have been given in kilometres, in which case the figures would have been 0.0, 1.0, 2.0, etc. making the table more concise. Pressure has been measured in megapascals, which is the SI pressure unit (1 MPa is approximately 10 atmospheres). Choice of unit can be difficult where an older but widely recognised unit remains in use and is more familiar. The density of the atmosphere is given in kilogrammes per cubic metre, abbreviated kg m^{-3}. An alternative abbreviation, kg/m^3, can be used but becomes confusing when mass, length and time units are involved together. A difficulty also arises when the numbers tabulated are either extremely small or extremely large. With mass there is little difficulty as we can usually convert to a conventional unit, e.g. a mass of 0.000 34 mg can be tabulated as 0.034 microgrammes (μg), but this is not always possible. For example large areas are not given a conventional prefix, so they are often

Table 4.1 The hypothetical state of the Earth's atmosphere at mid-latitudes (the standard atmosphere)

Altitude (m)	Pressure (MPa)	Temperature (°C)	Density (kg m⁻³)
0	0.1013	15.0	1.225
1000	0.898	8.5	1.1117
2000	0.795	2.0	1.0581
3000	0.701	−4.5	0.9093

tabulated by using a combination of **mantissa** and **exponent**. A value of 45 million km² could be written as 4.50 (mantissa) in the table with a header of 'area, 10^7 km²' (exponent of 10).

4.2 Graphs

There is a wide range of choice for graphical presentation, though the data themselves may dictate a single option. For samples containing ten or more measurements a **bar chart** or histogram is often used to show the structure of the data. For fewer measurements there is usually little point in constructing a frequency histogram. Bar charts tend to be used for discontinuous measurements where the only permitted values are integers, or they may describe lists of attributes. Bar charts can be used to plot quantities other than frequency on the y axis. For instance, information on the estimated wind power production in nine countries is shown in Figure 4.1. Discontinuous

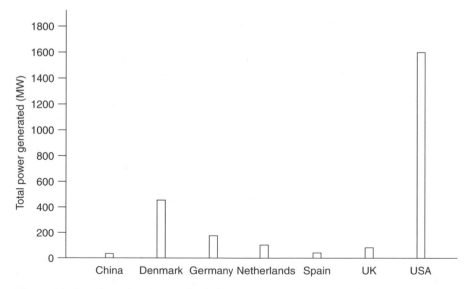

Figure 4.1 Bar chart showing total wind power generation in 1997 for seven countries.

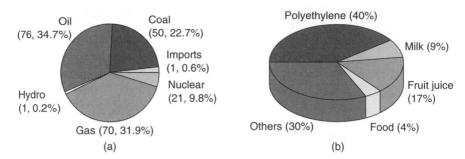

Figure 4.2 (a) Pie chart showing energy consumption as electricity in the UK for 1995; (b) a three-dimensional pie chart of plastic waste collected in a town on one day.

measurements are best displayed as bar charts where the individual bars are isolated from each other, reinforcing their separation into discrete entities. Continuous measurements should be plotted as a frequency histogram.

An alternative graphic for percentage frequencies is the *pie chart*, which is a circle divided into segments with the areas of the segments proportional to the percentage frequency (Figure 4.2). Pie charts have the advantage of displaying data in a concise, viewer-friendly fashion and allow much information to be presented on one page. To calculate the angle of the segment for a relative frequency of 23%, multiply 0.23 by 360 (number of degrees in a circle) to give 82.8°. The circle can then be segmented with the aid of a protractor. Colouring or infilling the segments can assist, providing the contrast is not made too extreme as this will overemphasise some frequencies. Sometimes 'three-dimensional' and 'exploded' pie charts are displayed by computer software, but there appears to be little advantage except some saving in space. They are difficult to draw without a computer and three-dimensional charts may not truly represent the data due to perspective.

Where circular measurements have been made involving angles or compass points, a circular graphic is advantageous. One of the most familiar is the *wind rose* used by meteorologists to indicate the frequency of wind directions. In Figure 4.3 a circle is divided into segments using the major compass points. The lines are drawn proportional to the frequency of winds blowing from the segment of 30° from whose mid-point the line is drawn.

A less familiar example is the *Maucha diagram* sometimes used by hydrologists to describe the ionic composition of a water sample (Figure 4.4). A circle is divided into eight equal segments and each segment is allotted a particular ion. On the left of the figure, the four commonest anions are shown, and on the right the four commonest cations. The area of each segment is made equal to the concentration of the ion in milliequivalents per litre so the completed diagram looks like an irregular-pointed star. Each diagram therefore depicts eight ion concentrations, facilitating comparisons between water types. The vertical separation of the cations and anions provides a useful check on ionic balance. If these eight ions make up the bulk of the

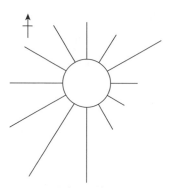

Figure 4.3 Example of a wind rose.

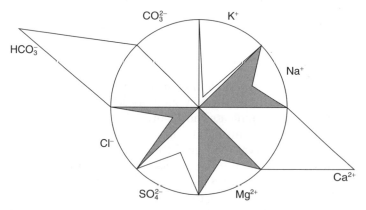

Figure 4.4 A Maucha diagram showing the ionic composition of world average freshwater.

composition, then the area on the left should be close to the area on the right. If it is not, then other ions must be significant, or the analysis has been performed incorrectly.

4.2.1 Frequency histograms

For continuous variables the frequency histogram is the preferred graphic, where the values along the x axis run continuously from left to right. The histogram is constructed by splitting up the measurements into a number of classes and then finding how many measurements fall in each class.

Suppose we have 50 measurements of windspeed taken with an anemometer, ranging from 0 to 18.5 km h^{-1}. To present the data as a frequency histogram the individual values need to be placed in classes. The class size, defined by the **class interval**, is usually kept constant. Choosing a class interval of 2 km h^{-1} would mean splitting the data into classes 0–1.9999 km h^{-1}, 2.0–3.9999 km h^{-1} and so on until all the data are included. The two

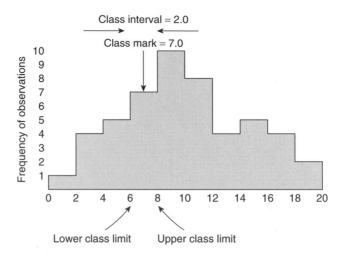

Figure 4.5 A frequency histogram showing the class interval, class mark and class limits.

figures defining the limits of a class are called the *lower class limit* and the *upper class limit,* respectively. The mid-point of each class is sometimes referred to as the *class mark,* as indicated in Figure 4.5.

Although the frequency histogram possesses a number of fixed features, the size of the class intervals will need to be decided. For small numbers of measurements ($n < 50$) the number of classes needs to be small; otherwise too many classes will contain no measurements at all and the histogram will not show the data structure clearly. The class interval is dependent upon the number of classes chosen and it is preferable to have an interval of unity, if possible, so that classes range 0–0.99, 1–1.99, etc. If this is not possible it is better to use an even-valued interval (e.g. 0–1.99, 2.0–3.99, etc.) than an odd-valued interval (e.g. 0–2.99, 3.0–5.99). There is no 'ideal' class number (N_c) for a given set of measurements but a rough estimate may be obtained as $N_c = 5 \log_{10} n$, where n is the total number of observations.

There are several modifications that can be made to the histogram to aid interpretation. One of these, used in demographic studies, is the population pyramid which shows the proportion of the population at different ages. As it is often instructive to separate on the basis of sex; the male and female histograms are drawn vertically and back to back as shown in Figure 4.6.

An alternative form of the histogram is the *dot plot* (Fig. 10.1, p. 86) where each datum occurring within a class interval is represented by a dot. This form is useful where the number of observations is less than 30 and a quick assessment of the data structure is required.

Occasionally you may be presented with data already given in unequal class sizes, or outliers may occur, with some values set well apart from the bulk. In these cases it is permissible to combine some of the classes. However, it is important to make sure that the block *area* rather than its height is made proportional to the frequency; otherwise the graph will be misleading. If area is used, then the *y* or **ordinate** axis can no longer be labelled

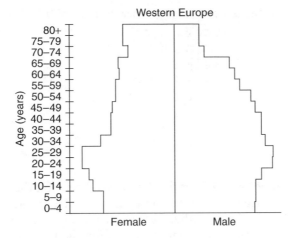

Figure 4.6 A population pyramid for Western Europe.

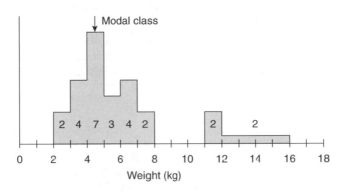

Figure 4.7 A frequency histogram with combined classes.

with the frequency, and each class will need to have the frequency of measurements printed within or above it (Figure 4.7). In the example in Figure 4.7 there were two measurements between 12.0 and 15.99 but their precise values were unknown, and the four classes covering this range have been combined into one. The area of this new block is then made equal to that of the preceding class (11.0–11.99), which also contains two measurements.

Although it is tempting to fill in the blocks of a frequency histogram with different colours or patterns of shading, this is best avoided as it is often distractive. Also ensure that the scale along the *x* or **abscissa** axis is continuous. There should not be any spaces between the frequency classes except for cases where the frequency for a particular class is zero; otherwise the histogram will resemble a bar chart.

4.2.2 Stem and leaf plots

The stem and leaf plot is a 'histogram–table' hybrid as it presents measurements in both a tabulated and a graphical form. An example of part of a stem and leaf plot is shown below for some rainwater pH values.

Stem and leaf plot for 35 pH measurements of 'acid rain'

```
3.6 | 2
3.7 | 3 5 8
3.8 | 1 4 5 9
3.9 | 1 2 2 3 8
4.0 | 0 2 3 4 4 5 7 7 8 9
4.1 | 0 1 3 5 5 9
4.2 | 4 5 7
4.3 | 2
4.4 | 6 8
```

The plot is prepared by ordering the data into a 'stem' on the left of the line and a 'leaf' on the right. There was only one pH value in the range 3.6–3.7, which was 3.62. The first two digits are shown in the stem and the third in the leaf. Where several measurements fall within the range of a stem value, e.g. there are three pH values between 3.7 and 3.8 (3.73, 3.75 and 3.78), they are placed in increasing size opposite 3.7, as shown in the example. If the measurements had been taken to a further significant figure, this is rounded to make the plot concise, e.g. a pH of 4.461 would be rounded to 4.46 and then plotted. The technique allows the data to be visualised and at the same time provides a ranking, which is useful if quantiles or non-parametric statistical tests are required.

4.2.3 Line graphs

A line graph is one method of illustrating how one variable changes with another. It is the most frequently employed graphic and can take several forms. A simple line graph is shown in Figure 4.8 demonstrating how the concentration of ammoniacal nitrogen in a sewage-contaminated river decreases downstream. The abscissa is used to display distance and the vertical ordinate the concentration. Note the small displacement of the y axis to permit a clear view of the first datum point.

Again modifications of the line graph are often made. One modification which is frequently used to display the changing abundance of plants or animals along an environmental gradient is the *kite diagram*. The line graph is drawn along the abscissa, which is used as a symmetry axis, and the line is repeated as a mirror image on the other side of the axis; the abscissa is then rotated 90°. An example showing seaweed zonation is given in Figure 4.9. The resulting enclosed area is often filled in as shown here, but this is not really necessary. Kite diagrams can be an effective and attractive means of displaying several abundancies along the same gradient, but unless a computer package is available they are time consuming to draw.

The choice of the axes of a line graph needs consideration and it is usual to display measurements of length, time or other independent variable (see Chapter 12) on the x axis. 'Time' takes precedence on the x axis and when it is used the graphic is called a *time series* (Figure 4.10). Line graphs should only be used on data that can be **interpolated**, i.e. at the interval or ratio level of measurement.

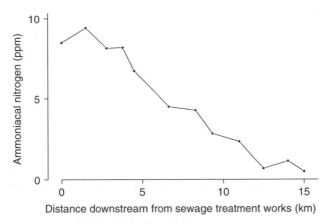

Figure 4.8 Line graph showing the decline of ammoniacal nitrogen in an effluent with distance downstream from a sewage treatment works.

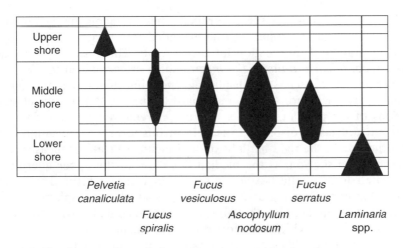

Figure 4.9 Kite diagram of seaweed zonation along a rocky shore.

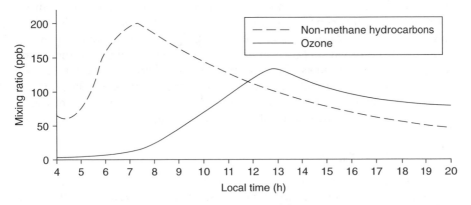

Figure 4.10 Time series showing mixing ratio profiles in a photochemical smog cycle.

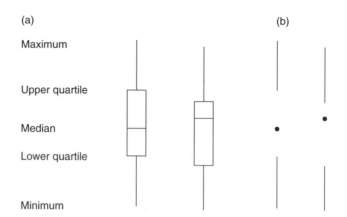

Figure 4.11 (a) Two boxplots showing the positions of the median, upper and lower quartiles and the range of observations for two samples; (b) abbreviated form of (a) providing the same information.

In cases where there are several ordinate measurements for each point on the x axis, the datum points can be extended into vertical lines or boxes providing statistical information for each sample. The most frequently used plots are *confidence limits to the mean*, which are explained in Chapter 9. Another effective means of displaying such information is the *boxplot* (Figure 4.11). In Figure 4.11(a) a standard boxplot is shown indicating the positions of the median, quartiles and the extreme range of the data. One of the advantages of displaying the data in this form is clarification of the data structure. In the right-hand boxplot of Figure 4.11(a), the measurements are seen to be skewed, as the median lies close to the upper quartile. An abbreviated form of the boxplot is shown in Figure 4.11(b). This form is suitable for hand-drawing as it is rapidly executed and displays the same information as the boxplot, though it might to be confused with confidence limits to the mean, which are plotted in a similar manner.

For publication quality graphics on A4 size paper, the line thickness should be at least 0.5 mm. Curved lines should not be drawn between the datum points unless there are good theoretical reasons for drawing a curved line. Hand-drawing a curve can be misleading and can lead to difficulties of **interpolation** between two points. If the data density of the graphic is high, there may be no need to draw lines between the points. Several independent sets of data can be shown on one graphic, using different symbols for the datum points linked by solid lines. If necessary, the right-hand ordinate can be used to show values of an additional variable. The graphic must not become too complex and should contain a maximum of five independent sets of measurements. The axes must be clearly labelled and the units of measurement shown.

4.2.4 Scattergraphs

When one variable is plotted against another and there are large numbers of points which do not follow a point-to-point obvious trend, a scattergraph

(a)

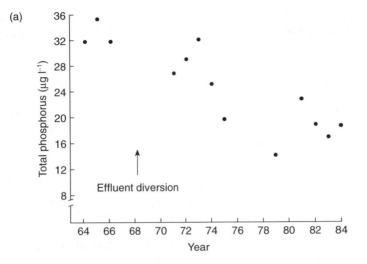

(b)

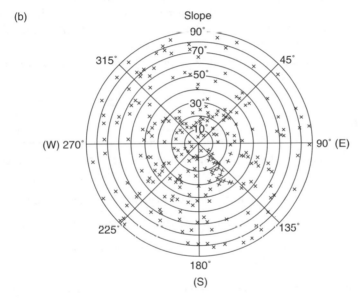

Figure 4.12 (a) Scattergraph showing the relationship between time and total phosphorus content of Lake Sammamish water following the diversion of sewage effluent. (With permission from Welch, E. B. *et al.*, 1986. *J. Wat. Pollut. Control. Fed.* 58: 92–6. Published by Routledge.) (b) A polar graph showing the positions of quadrats (×) on rock surfaces in Wales with reference to their compass positions and the slope of the surface (from Pentecost, 1979).

is often used to display the data. The points are left unjoined, though a **regression line** (see Chapter 12) may be drawn through the points in certain circumstances. Most scattergraphs display information on two axes at right angles (Figure 4.12(a)), but it is also possible to display data as a *polar graph* (Figure 4.12 (b)) or as a *trilinear graph*, where the axes are drawn at 60° to each other rather than the usual 90°. Three-dimensional graphs are available in many computer packages and can be useful to depict surfaces or points using three variables.

Producing a pie chart with Minitab

Data can be graphed in many ways using *Graph* on the menu bar. For example, to produce a pie chart you need two columns of a worksheet. In the first row of C1 put the frequency or percentage frequency and then type the attribute name in the first row of C2, e.g.

C1	C2
22.5	gas
18.7	oil
58.8	coal

Now click on *Graph > Pie chart*. In the dialog box click on 'Chart table'. Enter C2 in 'Categories' and C1 in 'Frequencies in'.

If you want the data in the pie chart to be presented in the same order as that shown in the worksheet, click against 'Worksheet' in the 'Order of Categories'. Now click 'OK' to see the pie chart.

You do not need to provide data as percentage frequencies; Minitab will calculate this for you.

Key notes

- Tabulated data need to be self-explanatory with recognised symbols, units and quantities.

- Bar or pie charts should be used to display the frequencies of discontinuous measurements.

- Histograms are used to display the frequencies of continuous measurements.

- Stem and leaf plots can be used to display continuous measurements in a combined tabulated and graphical form.

- Line graphs and scattergraphs show how one variable changes as a function of another variable.

CHAPTER FIVE

The normal curve and inferential statistics

The frequency histogram shown in Figure 5.1(a) resembles a section through a bell – a shape frequently encountered with large-interval data sets. The smooth curve on the right (Figure 5.1(b)) shows a 'normal curve' obtained by plotting the following mathematical expression

$$f(x) = \frac{1}{\sigma\sqrt{(2\pi)}}\exp\left[-\frac{(x - \mu)^2}{2\sigma^2}\right]$$

The expression contains two important numbers, which describe the position of the curve on the x axis and its shape. These numbers are called **parameters**. The first, given the symbol μ, is also known as the *population mean* and the second (σ) the *population standard deviation*.

The similarity between these curves, one obtained from theory and the other from sampling, was appreciated as early as the eighteenth century. Even earlier, the astronomer Galileo reasoned that instrumental observations were liable to error, with the error equally likely to fall above and below the true value and with most of the observations clustering around the true value. If the 'true' value could be obtained, then the frequency distribution of the errors would resemble the 'normal' curve, with the true value located at the mean. The observation that instrumental error is equally likely to fall above or below the true value has been exploited in some interesting ways. For example, when the height of Mount Everest was first determined in the nineteenth century, levelling had to be carried out over hundreds of kilometres of countryside to the base of the mountain. A large number of individual levellings had to be made. As each level had a small error associated with it, it was reasoned that the large number of errors would cancel, leading to a good estimate. Modern methods of measurement have demonstrated the truth of this statement.

The normal curve was first expressed on paper by A. de Moivre in 1733. Applications of the 'normal law' were slow to come, and originated

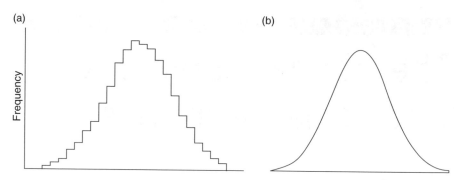

Figure 5.1 (a) Frequency distribution from a series of measurements approximating the normal distribution; (b) a normal curve plotted with a mathematical function.

in astronomy. Biological applications came later. In the mid-nineteenth century the Belgian statistician Adolphe Quetelet obtained a good fit between a form of the normal curve and the height of 100 000 conscripts. Modern parametric statistics employing the properties of the normal curve were not applied widely to environmental problems until the 1950s. Much of the theory underlying the methods described in this book, particularly those of **non-parametric** statistics, were developed in the twentieth century.

The term 'normal law' was first used by the American C. S. Peirce in 1873. It had earlier been termed the 'law of frequency of error', among others. The adjective 'normal' has sometimes been objected to, as the distributions of many natural phenomena do not follow the normal curve. Use of the term 'normal' is too widely established, however, to allow a change.

We shall see below how properties of the normal curve can be used to make some simple statements about certain kinds of measurements. This will lead to an understanding of *inferential* statistics.

5.1 The standard normal curve

While the important properties of the normal curve remain the same whatever the values of its parameters, it is useful to consider a standard normal curve whose population mean (μ) is zero and with the total area under the curve equal to one (Figure 5.2(a)). Units along the x axis are measured in standard deviation units and you can see from the figure that a distance of two standard deviations from the mean covers most of the area under the normal curve. Since the form of the curve is fixed, with the population mean μ being equal to the mode (the summit of the curve), it is possible to make some useful statements about the curve.

The area enclosing one standard deviation on either side of the mean of a standard normal curve as shown in Figure 5.2(a) is fixed at 0.6826 (68.26%) of the total area. Likewise in Figure 5.2(b) the area enclosing ±2 standard deviation units from the mean is 0.9544 (95.44%). In Figure 5.2(c) the area

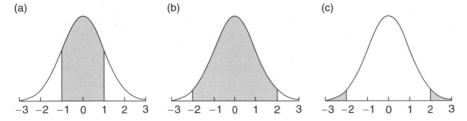

Figure 5.2 A series of standard normal curves: (a) the shaded area lies within one standard deviation of the standardised mean (0); (b) the shaded area is within two standard deviations of the mean; (c) the shaded area lies outside two standard deviations from the mean.

in the tails (extremities) is shaded. Since the entire area of the curve is one unit of area (100%), this required area ($\pm 2\sigma$) must be $100 - 95.44 = 4.56\%$. Note that, as the curve is symmetric about the mean, the area–standard deviation relationships below the mean (deemed negative, since the mean is zero) are the same as those above the mean. If areas corresponding to other standard deviations or their fractions are required, they can be found with reference to tables of the standard normal curve (**Table II**).

The relationship between standard deviation and probability has some useful practical applications, illustrated in Box 5.1.

BOX 5.1: PRACTICAL APPLICATION OF THE NORMAL CURVE

Suppose that a large sample of normally distributed suspended solids determinations is available for a particular site on a river. The sample mean ($\bar{x}$) is found to be 18.3 ppm and the sample standard deviation (s) is 8.2 ppm. We are asked to find the probability of a random sample containing 30 ppm suspended solids or more. This information is sketched onto an appropriate normal curve below and the area (probability) required is shaded.

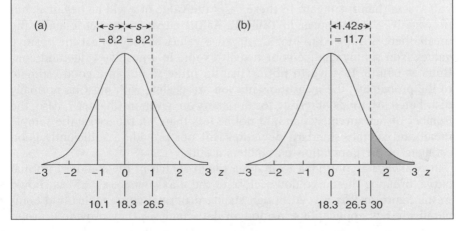

We can obtain the probability of obtaining a random measurement with suspended solids greater than 30 ppm with reference to the table of areas under the standard normal curve. The mean must first be transformed from the sample value (18.3) to zero, and then the difference between the stated value (30) and the mean is converted into units of the sample standard deviation (see the figure) so that the probability can be read from **Table II**.

The value of 30 ppm can be changed into standard deviation units from the mean by subtracting the sample mean from 30 ppm and dividing the difference by the sample standard deviation. This quantity is termed the *standardised deviation* or z value. In this case

$$z = (30 - \bar{x})/s$$

giving

$$z = (30 - 18.3)/8.2 = 1.4268$$

Thus, on the standard normal curve sketched in the figure, the value corresponding to 30 ppm is equal to 1.4268 standard deviation units above the mean. If you turn to **Table II** you will see values of z in the left-hand column. To find the area corresponding to $z = 1.43$, go down the column until you find $z = 1.4$. To find 1.43, work across the row until you come to the column headed '0.3'; the area value is 0.4236.

Note in the sketch above the table that the areas given refer to those *from* the mean to the value of z, whereas we need a value *from* z into the upper tail of the standard normal curve, i.e. we are asked for the probability of a value greater than 30 ppm. We can find this area because we know that the area under the entire standard normal curve is 1.0. Therefore the area under the upper half of the curve, above the mean, must be 0.5 because all normal curves are symmetric about the mean. The area we require, shaded in the figure, is therefore $0.5 - 0.4236 = 0.0764$ or 7.64%. This means that in a large number (say 1000) of random samplings the suspended solids value will be 30 ppm or more on about 76 occasions.

You might also need to know how to use the standard normal curve for values less than the mean. In these cases the value of z will be negative, but no negative values appear in **Table II**. As the curve is symmetric about the mean, there is no need to include negative values as they repeat the positive values. You simply change your negative value to a positive value and continue as before. It is worth noting that, in order to obtain a good estimate of the probability, the measurements you are dealing with must be normally distributed. Methods of testing for normality are given in Chapter 7. Also, the number of measurements should not be less than 30; otherwise the sample mean and sample standard deviation will not provide a sufficiently good estimate of the population parameters μ and σ.

It is always a good idea to make a rough sketch of the standard normal curve, bringing the curve down close to the x axis where $z = \pm 3$, as shown in the figure in Box 5.1. Although standard normal curve tables find comparatively few applications in modern data analysis, they provide a useful exercise in simple probability calculations using the normal distribution and a practical introduction to inferential statistics.

5.2 Central limit theorem

With an extremely large set of measurements we should have little trouble in deciding whether or not the normal distribution provides a good fit to our data. We could also divide up the measurements into a series of smaller subsets and make some interesting discoveries. Suppose that we have a large sample consisting of 10 000 interval measurements. We produce 1000 subsets each containing ten measurements taken at random from the large sample in Figure 5.3.

If the mean of each of the subsets is calculated, we end up with 1000 means, still quite a large sample. If we now treat this collection of means as a *new* sample with $n = 1000$, we can plot another frequency histogram, this time of the means. It should come as no surprise that the new histogram differs from the original containing all 10 000 individual measurements. We would not expect to see such a wide variation in the 'means' histogram since, by taking means of ten samples at a time, we smooth out the extreme values within the sets. Thus, there is less variation in the 'means' sample. The standard deviation will be smaller and there is less 'spread' in the histogram. This is clearly seen in Figure 5.3. Another feature of the means histogram is also worth noting. The frequency distribution for the means appears more symmetric than that of the original data.

We can also take our sample of 10 000 measurements and divide it into 100 subsets each containing 100 values. If the means of each of these larger subsamples are plotted, we find that the standard deviation has reduced even more and the frequency distribution is very close to the normal distribution. If an even bigger sample had been taken for averaging, we would find that the standard deviation would keep falling in value and the normal distribution would fit the data even better. The phenomenon is embodied in the *central limit theorem*, which states that:

As sample size increases, the means of samples drawn from a population of any distribution will approach the normal distribution.

Particular use of the central limit theorem is made in calculating the 'standard deviation of the mean' and in the comparison of sample means (Chapters 8 and 9).

5.3 The standard deviation of the sample mean

Mathematicians have calculated the 'expected' value of the standard deviation of the mean as a function of the sample size (n) and the sample standard deviation (s). The relationship turns out to be surprisingly simple. The sample standard deviation of the mean ($s_{\bar{x}}$) is given by

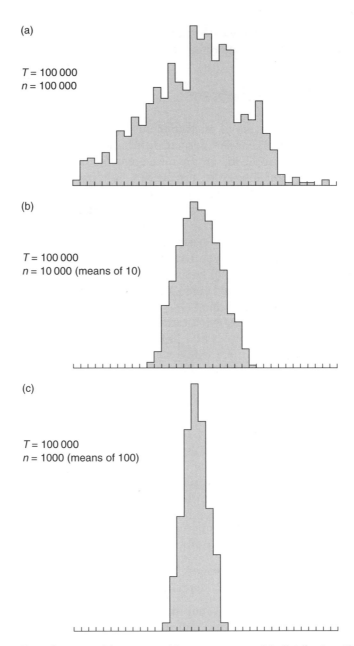

Figure 5.3 Three frequency histograms with common axes: (a) distribution of a large raw data sample; (b) distribution of means based on subsamples of 10 randomly drawn measurements; (c) distribution of means based on subsamples of 100 randomly drawn measurements. T is the total number of original observations.

$$s_{\bar{x}} = s/\sqrt{n}$$

Therefore the standard deviation of the mean can be estimated from the standard deviation of the sample. It is not necessary to use huge numbers of measurements and obtain a set of means for the calculation. The form of this equation, with the sample standard deviation divided by the square root of the number of measurements, ensures that, as n increases, the standard deviation of the mean gets smaller, as indicated in Figure 5.3. The standard deviation of the mean ($s_{\bar{x}}$) is sometimes referred to as the *standard error of the mean* or simply as the *standard error*. The former term is preferred, as standard deviations can be obtained for many other statistics, such as the median, and even the sample standard deviation itself, which could lead to confusion.

5.4 Testing hypotheses

In statistical analysis, hypothesis testing is employed so that meaningful statements can be made about the populations from which samples are drawn. The hypothesis under test is called the *null hypothesis*. In research, measurements are frequently made to test a prediction. In a statistical analysis, the null hypothesis will nullify this prediction – hence the name. Significance testing is an integral part of hypothesis testing and the two procedures are best introduced with a simple example. Suppose a single observation (z) is made from a population represented by the standard normal curve (Figure 5.4). The hypothesis under test, the null hypothesis, will be that z has been chosen at random from the population represented by the curve. It should be apparent from the curve that frequencies of z values close to the mean ($\mu = 0$) are high, since the mean corresponds to the mode, while frequencies away from the mean decline. Beyond a distance of three standard deviations from the mean the frequencies become very small indeed. In Figure 5.4, two values of z are shown, $z = 1.96$ and $z = 2.58$. For each of these values, using **Table II** we can obtain the probability of z values *greater* than

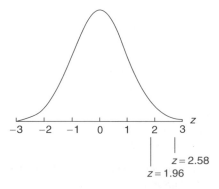

Figure 5.4 Standard normal curve indicating two randomly obtained values of z.

these figures as areas under the curve. For $z = 1.96$, the area of the curve to the right of the z value is $0.5 - 0.475 = 0.025$ or 2.5%. This area is equivalent to the probability of finding a z value greater than 1.96 chosen at random from the population. Likewise, for $z = 2.58$, the probability of a greater value is $0.5 - 0.495 = 0.005$ or 0.5%. Since the curve is symmetrical about the mean, the probability of obtaining a value of z less than -1.96 is also 2.5%, so the total probability of obtaining a z value which is either less than -1.96 or greater than $+1.96$ is $2.5 + 2.5 = 5$%. Since the total area of this normal curve is unity (100%), the probability of obtaining a value of z *between* -1.96 and $+1.96$ is 95%. Likewise between $z = \pm 2.58$ it is 99%. We can now frame the null hypothesis in terms of the probability by stating that a random observation of the population will have a value between -1.96 and $+1.96$ with 95% probability. Now consider the situation where the above hypothesis has been stated and we obtain a random observation of z which lies *outside* these limits. There are two possibilities. Either we have chosen an 'unlikely' value of z, or our hypothesis is incorrect. In the latter case, it may be that our value of z is in fact quite a 'likely' value but it has been drawn from a different, unknown population with a mean $\mu \neq 0$. When performing a significance test, we make the rule that if our value of z lies outside the range ± 1.96 the null hypothesis is *rejected* and the value of z is termed *significant at the 5% level*. Alternatively, we state that the probability (p) that the observation comes from a normal distribution with $\mu = 0$ is less than 5% (often abbreviated as $p < 5$% or $p < 0.05$). In the case where $z = 2.58$ the observation is even less likely to occur (about one in 100 observations) and the value is termed *significant at the 1% level*. As a general rule, the cut-off point for significance tests is the 5% level.

An interesting analogy exists between hypothesis testing and court judgements where the accused is presumed innocent until proved guilty beyond reasonable doubt. Here the null hypothesis is innocence, and if reasonable proof to the contrary is admitted, the null hypothesis is rejected and the alternative, guilt, is accepted. Experience shows that the innocent are sometimes guilty and the guilty are sometimes acquitted. The same is true in statistical tests. Our test result can only attach a certain probability to our conclusions. We can never be absolutely sure that we have come to the correct conclusion.

Key notes

- Frequency histograms of interval and ratio measurements often approximate the normal distribution.

- The normal distribution is a mathematical curve whose shape and location is determined by two population parameters.

- The population parameters of the normal curve are the population mean (μ) and the population standard deviation (σ).

- Areas beneath the standard normal curve ($\mu = 0$, $\sigma = 1$) correspond to the probability of occurrence of normally distributed measurements with specified values.

- The probability of occurrence of specified measurements can be estimated using tabulated values of the standardised deviate (z).

- The central limit theorem states that the sample mean ($\bar{x}$) is a normally distributed quantity.

- Significance testing involves setting up a null hypothesis and then providing evidence for its acceptance or rejection. If the null hypothesis is rejected on the statistical evidence, the alternative hypothesis must be accepted.

- In terms of probability of occurrence, the null hypothesis is rejected if its probability value obtained from statistical tables is less than 5%. In such cases there is less than a one in 20 chance that the null hypothesis is correct.

CHAPTER SIX

Frequency distributions

In Chapter 3 methods were introduced to measure the frequency of observations. The nominal or lowest level of measurement was introduced first and in Chapter 6 some methods are shown for the analysis of such data. Most of the methods employ cross-tabulations, often referred to as 'contingency tables', where the data, consisting of the nominal frequencies, are written down in *cells* of a cross-tabulation. This is simple to do and allows the data to be efficiently described and analysed. Cross-tabulations are often used to analyse frequency data of the following types: male/female; agree/disagree; or present/absent. These are examples of dichotomous classifications where there are only two states for each variable (attribute). For instance, we may wish to analyse a questionnaire where males and females either agree or disagree with a particular statement. There can only be four outcomes, namely (1) males agree, (2) males disagree, (3) females agree, (4) females disagree. If data were collected from a total of 100 questionnaires we might find 12 agreeing males, 27 disagreeing males, 30 agreeing females and 31 disagreeing females. These four figures could be entered into a cross-tabulation as shown in Table 6.1.

To clarify the information, it is helpful to include the total number of responding males and females. In a questionnaire, and indeed in many experiments, there is no reason to suppose that there will be equal numbers of males and females. In addition, it would be useful to know the total number of people who agreed and disagreed. This information can be added to the cross-tabulation (Table 6.2). The sums are known as *marginal totals*. The total number of observations (n) is shown lower right.

Before analysing this cross-tabulation, it is worth pointing out that variables with more than two categories can be treated in the same way. With large numbers of categories, however, it becomes more difficult to interpret the results and more involved techniques are required. As mentioned in Chapter 16, it is also possible to reduce an ordinal or interval level of

Table 6.1 Example of a 2 × 2 cross-tabulation: males and females agreeing or disagreeing with a question

	Males	Females
Agree	12	30
Disagree	27	31

Table 6.2 Cross-tabulation with marginal totals (*n*)

	Males	Females	*n*
Agree	12	30	*42*
Disagree	27	31	*58*
n	*39*	*61*	**100**

measurement to the nominal level. For instance, a sample consisting of four ages, 24, 65, 75, 23 years, can be allocated into age classes of 20–29, 30–39, 40–49 and so on and then presented in the cross-tabulation. It is also important to note that only two variables (or attributes) are normally investigated in a cross-tabulation, though methods also exist for multivariate analyses.

The main reason for preparing and analysing a cross-tabulation is to test for independence between the attributes. In the cross-tabulation in Table 6.2 we would like to know if the answers given are independent of sex. If this is the case the same proportion of males and females will agree with the statement. If the proportion differs, with the females agreeing more often than the males, for example, it could be argued that the difference is merely the result of chance. However, if we consistently obtained the same result, or if the data set was large, we would suspect that a real difference between the males and females existed. There are a number of simple statistical tests that can help us decide.

6.1 The χ^2 test

The χ^2 test can be used to test the null hypotheses that the observations (not the variables) are independent of each other for the population. The test is based upon the difference between the observed frequencies shown in the cells of the cross-tabulation and the *expected* frequencies that would be obtained if the variables were truly independent. If the null hypothesis is true then both the difference between the observed and expected frequencies and the value of χ^2 will be small.

$$\chi^2 = \sum \frac{(\text{observed frequency} - \text{expected frequency})^2}{\text{expected frequency}}$$

To complete the χ^2 test, the expected frequencies need to be calculated. They are included in the following equation, which is based on a 2×2 cross-tabulation.

		Variable A		
		State 1	State 2	
Variable B	State 1	a	b	$a + b$
	State 2	c	d	$c + d$
		$a + c$	$b + d$	$n = a + b + c + d$

$$\chi^2 = \frac{n(ad - bc)^2}{(a + b)(c + d)(a + c)(b + d)}$$

To complete the test, χ^2 tables need to be consulted using appropriate **degrees of freedom**. Degrees of freedom are integer numbers, which have to be calculated for many statistical tests and involve a simple subtraction. They arise from a consideration of the number of independent units present in a sample after the calculation of statistics. The 2×2 table provides a simple demonstration of this. With fixed marginal totals, once a value is assigned to any one of the four cells, the remaining three values are fixed by the constraint that they must add to give the marginal values. There is therefore only one independent unit, i.e. one degree of freedom. A general equation which can be used to calculate degrees of freedom in the χ^2 test for any table size is

$$df = (r - 1)(c - 1)$$

where r stands for the number of rows of cells and c the number of columns of cells. For a 2×2 table, such as that shown above, $r = 2$ and $c = 2$, so the degrees of freedom are $(2 - 1)(2 - 1) = 1^2 = 1$.

BOX 6.1: A χ^2 TEST FOR A 2×2 CROSS-TABULATION

The χ^2 test is applied to the questionnaire example above, where $a = 12$, $b = 30$, $c = 27$ and $d = 31$.

$$\chi^2 = \frac{100(12 \times 31 - 30 \times 27)^2}{(12 + 30)(27 + 31)(12 + 27)(30 + 31)}$$

$$= \frac{100 \times (372 - 810)^2}{42 \times 58 \times 39 \times 61}$$

$$= \frac{100 \times (-438)^2}{5\ 795\ 244}$$

$$= 3.310$$

Turn to the critical values of χ^2 (**Table VI**), a small portion of which is shown below.

P	5	1	0.1
$v = 1$	3.84	6.63	10.83
2	5.99	9.21	13.81
3	7.81	11.34	16.27
4	9.49	13.28	18.47
5	11.07	15.09	20.52
6	12.59	16.81	22.46

Note that the values are presented in three columns for significance levels in the range 5%–0.1% against the degrees of freedom. As with the normal distribution, unlikely values of χ^2 are found in the tails of the distribution. Using the 5% (0.05) level, the critical value of χ^2 for one degree of freedom is **3.84**. This means that only 5% of the area under the χ^2 curve lies beyond the value 3.84 on the x axis. Our test value does not exceed this critical value and so the null hypothesis is accepted. There is no reason to believe that males and females differ with respect to the question asked – the differences observed most probably arose by chance.

6.2 Yates' continuity correction

The χ^2 distribution differs from the normal distribution and the shape of the distribution depends on the degrees of freedom. It is based upon the distribution of a population of sums of squares, which makes it asymmetric about the median (compare the t distribution, Chapter 8). χ^2 is also a continuous distribution, while the frequencies being analysed are discontinuous, since the figures entered into the 'observed' cells will be whole numbers. To improve the analysis, Yates' correction is often applied. This involves modifying the χ^2 equation for the 2×2 table in the following way

$$\chi^2 = \frac{n\left(\left|\, ad - bc \,\right| - 0.5n\right)^2}{(a+b)(c+d)(a+c)(b+d)}$$

where the quantity within the vertical bars is the **modulus** (i.e. the positive value of the calculated quantity). Applying this to the above example

$$\chi^2 = \frac{100\left(\left|\, 12 \times 31 - 30 \times 27 \,\right| - 0.5 \times 100\right)^2}{5\,795\,244}$$

$$= \frac{100\left(\left|\, 372 - 810 \,\right| - 50\right)^2}{5\,795\,244}$$

$$= \frac{100 \times (388)^2}{5\,795\,244}$$

$$= 2.598$$

Comparing the corrected value with the critical value of χ^2 (**Table VI**) shows that the test value is lower than the critical value (**3.84**) and the null hypothesis is accepted, as before. Recent work has shown that Yates' correction usually overcompensates and some researchers no longer apply it, particularly in cases where $n > 200$.

To be used correctly, the χ^2 test must be used on cross-tabulations where fewer than 20% of the 'expected' frequencies are less than five and none of the expected frequencies is less than one. If this does not hold, the problem can sometimes be overcome by combining categories.

6.3 The *G* test

The *G* test is similar to the χ^2 test and may be considered as an alternative test of independence of observations in cross-tabulations. Many statisticians believe that the test is superior to the χ^2 test, particularly in more complex situations, although at present it is not as popular. In common with the χ^2 test, observed and expected frequencies are obtained as shown in Box 6.2.

BOX 6.2: EXAMPLE OF A *G* TEST

Data from the questionnaire example above will be used. First it will be necessary to calculate the 'expected' frequencies using another method.

	Males	Females
Agree	*a*	*b*
Disagree	*c*	*d*

The 'expected' frequency for cell *a* ($f_{expected-a}$) is obtained as

$$f_{expected-a} = (a + b)(a + c)/n$$

where *n* is the total number of observations. To obtain the expected frequency of any cell, you add the observed frequency of that cell to the other observed frequency in the same *row*. This quantity is multiplied by the frequency of the desired cell added to the other cell in the same *column*. The whole is then divided by *n*. For the example, the following results are obtained, with the 'expected' frequencies shown in italics. The expected frequencies are fractional, which could not be obtained from an actual experiment, but the decimal points are retained to improve accuracy.

	Males		Females	
Agree	12	*16.38*	30	*25.62*
Disagree	27	*22.62*	31	*35.38*
				n = 100

The next quantity to calculate is

$$G = 2\sum O \ln(O/E)$$

where O and E are the observed and expected frequencies, respectively, in each cell; 'ln' is the natural logarithm, which is obtained using a function key of a scientific calculator (see Appendix 3).

First, add up the four quantities $O \ln(O/E)$ as follows

$$12 \ln(12/16.38) + 30 \ln(30/25.62) + 27 \ln(27/22.62) + 31 \ln(31/35.38)$$

$$= 12(-0.3112) + 30(0.1578) + 27(0.1770) + 31(-0.1322)$$

$$= -3.734 + 4.734 + 4.779 - 4.098$$

$$= 1.681$$

Care must be taken adding the positive and negative numbers. Now multiply the above sum by two to obtain G

$$G = 2 \times 1.681 = 3.362$$

Finally, the value of G is divided by *William's correction*

$$1 + [(a^2 - 1)/6nd]$$

where a is the number of frequency cells, n is the total number of measurements and d is the number of degrees of freedom $(r - 1)(c - 1) = 1$. For this example, $a = 4$, $n = 100$ and $d = (2 - 1)(2 - 1)$, giving William's correction as

$$1 + (16 - 1)/6 \times 100 \times 1 = 1 + 15/600 = 1.025$$

Dividing G by this value gives G_{adj} (adjusted) $= 3.362/3.28 = 1.426$. This value of G_{adj} is compared with critical values of χ^2 (**Table VI**) with one degree of freedom (**3.84**), which once again leads to acceptance of the null hypothesis.

For 2×2 tables these tests will be found satisfactory for analysing a wide range of data, but better methods are available in particular circumstances. In cases where the expected frequencies are less than about five, *Fisher's exact test* is preferable. If the data are matched or paired, then *McNemar's test* for correlated proportions, also based on χ^2, can be used. Methods are also available to combine several individual 2×2 tests taken from different experiments or surveys. Details of these methods can be found in Everitt (1977) and Sokal and Rohlf (1995).

6.4 ϕ and related methods of association

The calculated χ^2 value can be modified to provide a new statistic, ϕ, with some useful applications.

$$\phi = \sqrt{(\chi^2/n)}$$

ϕ takes values from zero to ± 1 and is called a coefficient of association. The sign of ϕ is obtained from $ad - bc$ in the cross-tabulation and demonstrates either a negative or a positive association between the two criteria under test. Strong associations occur when the value of ϕ approaches ± 1. It is therefore similar to the correlation coefficient for bivariate data (Chapter 11).

For the questionnaire example

$$\phi = \sqrt{(2.598/100)} = 0.161$$

Since $ad - bc = -438$, the association is both small and negative. The statistic is useful for comparing the results of several χ^2 analyses. There are other comparative measures, such as Cramer's V for larger tables and the 'proportional reduction of error' (PRE). Further discussion and applications of these methods can be found in Leach (1979) and de Vaus (1996).

6.5 Other cross-tabulations

Similar methods are available for analysing the following forms of table, where f represents a frequency of observations:

(a) $f_1 \, f_2 \, f_3 \, f_4 \, f_5 \, f_6 \, f_7$ (b) $f_1 \, f_2$

(c) $f_1 \, f_2 \, f_3 \, f_4 \, f_5 \, f_6$ (d) $f_1 \, f_2 \, f_3$
$\quad\;\; f_7 \, f_8 \, f_9 \, f_{10} \, f_{11} \, f_{12}$ $f_4 \, f_5 \, f_6$
$\qquad\qquad\qquad\qquad\qquad\qquad\quad f_7 \, f_8 \, f_9$

Example (a) is a *string*, which might represent the frequency of seven species of plant present in 165 quadrats. The 'expected' frequencies are calculated as the total of the frequencies divided by the number of categories. In the above case, with seven categories, the expected values are all 165/7 or 23.57. χ^2 can then be calculated using $\Sigma[(\text{observed} - \text{expected})^2/\text{expected}]$ with $i - 1$ degrees of freedom, where i is the number of categories, giving $7 - 1 = 6$ in this example. This form of the test finds use in goodness-of-fit studies, where an observed frequency distribution is compared with an expected distribution. A situation which often occurs in questionnaire analysis is the one-sample case, where there are only two categories (b). For instance a simple yes–no response to a question. Out of 100 responses, 41 might reply 'yes' and 59 reply 'no'. The expected values will be 50 in each case and χ^2 is obtained as

$$(41 - 50)^2/50 + (59 - 50)^2/50 = 3.24$$

Ideally, a continuity correction should be applied, which will slightly reduce this estimate of χ^2.

Example (c) is a $2 \times c$ table and is one form of the general $r \times c$ table (row $\times$ column) shown in example (d). These tables are more difficult to analyse and interpret compared with the 2×2 table, but the techniques are basically similar.

χ^2 using Minitab
The frequencies are first entered into a worksheet as follows (a 3×3 example):

	column		
	1	*2*	*3*
row *1*	38	45	53
2	62	69	74
3	47	98	13

Stat > tables > Chi-Square Test

Under 'Columns containing the table' type C1–C3. Click 'OK'.

You should receive the following output giving the χ^2 value and its probability of occurrence under the null hypothesis, H_0. In this example, the null hypothesis will be rejected ($p = 0.000$) showing that there is significant association between one or more of the 3×3 categories. The probability is in fact a finite value and should really be given as ($p \le 0.00005$), but Minitab has rounded the value to zero.

```
Expected counts are printed below observed counts
              C1               C2               C3            Total
  1           38               45               53             136
              40.06            57.78            38.16
  2           62               69               74             205
              60.39            87.09            57.52
  3           47               98               13             158
              46.55            67.13            44.33
Total        147              212              140             499
Chi-Sq = 0.106 + 2.827 + 5.775 +
         0.043 + 3.759 + 4.725 +
         0.004 + 14.200 + 22.141 = 53.580
DF = 4, P-Value = 0.000
```

Minitab allows blocks of data from 2×2 to 7×7 to be analysed.

Key notes

- Frequency analysis is applied to observations made at the nominal level.

- Data are displayed as cross-tabulations for the analysis.

- Tests of independence of the observations are performed using χ^2 or the G statistic.

Exercises

6.1 In a questionnaire, 259 adults were asked what they thought about tax increases on fuel as a means of popularising public transport. 113 people agreed that increasing the tax would be beneficial, while the remainder disagreed. Perform a χ^2 test to determine the probability of the results being obtained by chance.

6.2 In another questionnaire, men and women were asked if they supported or opposed a ban on smoking in public places. The results are shown in the cross-tabulation below. Perform a χ^2 test to obtain the probability of these results arising by chance.

	Men	Women
Oppose	58	63
Support	57	45

6.3 Small boat owners were asked if they had ever used antifouling paints to protect their hulls from plant and animal growth, and if their boat exceeded a length of 5 m. Perform a χ^2 test of independence.

	Used antifouling paint	
	Yes	No
Boat <5 m long	18	43
Boat >5 m long	26	9

6.4 As part of a marshland survey, an investigator needed to know whether two threatened marsh plants, *Parnassia palustris* and *Trollius europaeus*, were associated in her 1×1 m quadrats. Use the *G* test to test this hypothesis.

		Parnassia	
		Present	Absent
Trollius	Present	54	13
	Absent	45	31

6.5 In a study of salt damage to town walls a survey was undertaken to see whether walls built of sandstone were more prone to salt damage than walls of limestone. Damage was assessed by recording the presence/absence of exfoliation (peeling of the surface). Use a *G* test to analyse the data.

	Limestone	Sandstone
Exfoliation	13	58
No exfoliation	46	62

6.6 In a roadside survey, private car drivers were asked which type of vehicle fuel they used. The results were as follows: 78 used unleaded fuel, 25 used super unleaded, 65 used leaded and 69 diesel. Use a χ^2 test to determine the probability of the figures arising by chance.

CHAPTER SEVEN

Recognising the normal distribution

Providing you have a moderately large interval data set ($n > 15$) it is possible to conduct some simple tests for a normal distribution. Recognising the distribution of your data is important as it provides a firm base on which to establish and test hypotheses. If the data provide a reasonable fit to the normal distribution, then they may be analysed and compared with other normally distributed data in the literature. If the data are not normally distributed, some of the normality tests can indicate an alternative model or a mathematical transformation with which it is possible to convert the data to the normal distribution. This can often prove useful as it allows for a more thorough analysis. Four methods for testing for a normal distribution are given. The last, a 'goodness-of-fit' test, can be extended to many other types of distribution.

7.1 Inspection of the frequency histogram

The frequency histogram is drawn and the mean, median, mode and standard deviation are obtained. If the first three statistics occur in the same class interval and virtually all of the measurements fall within two standard deviation units on the two sides of the mean, then a normal distribution may be assumed. For the distributions in Figure 7.1, that on the left approximates the normal distribution while that on the right does not. A stem and leaf plot can also be used to observe the distribution.

7.2 Use of rankit tables

If the number of measurements lies between ten and 30 the rankit graphical method can be used. The rankit values are the mean positions in standard deviation units of ranked items taken from a standard normal distribution with $\mu = 0$ and $\sigma = 1$. The method is illustrated in Box 7.1.

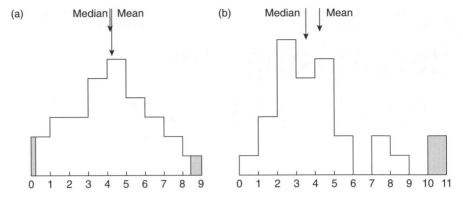

Figure 7.1 Two frequency distributions showing the positions of the mean and two standard deviations from the mean. The shaded areas lie outside two standard deviations from the mean. (a) An approximately normal distribution; (b) a skewed (non-normal) distribution.

BOX 7.1: TESTING FOR NORMALITY USING RANKITS

Measurements of quarry dust, which number 15 in this example, are first ranked in increasing order as shown in column 1.

Particle diameter (μm) in quarry dust

1 Ranked measurements	2 Rankit for n = 15	3 Rankit corrected for ties
19.9	−1.736	−1.736
22.6	−1.248	−1.248
25.0	−0.948	−0.948
26.9	−0.715	−0.715
30.5	−0.516	−0.516
31.7	−0.335	−0.335
31.9	−0.165	−0.165
32.2	0.000	0.000
33.2	0.165	0.165
33.9	0.335	0.426
33.9	0.516	0.426
35.6	0.715	0.715
36.2	0.948	0.948
38.5	1.248	1.248
44.1	1.736	1.736

Next, turn to **Table I** and locate the rankits for 'size of sample 15'. A column of eight numbers will be found ending with the value 0.000. It is important to note that rankits are only given up to the number corresponding to the median of the measurements, which in this case is the value 32.2 in column 1. Once the median is reached, the rankits are repeated in the table but with a change in sign.

The lowest of the ranked measurements (19.9) is given the lowest-valued rankit. This is not the rankit value 0.000, as might be expected, but the highest value in the rankit table given a negative sign. The lowest rankit is therefore −1.736. The next lowest measurement (22.6) is given the next lowest rankit (−1.248) and so on until the median is reached. The rankits are then reckoned positive so that the largest value of the measurements receives the largest rankit (+1.736).

When the sample size is odd, as in the case above, the median is always given the rankit 0.000, as can be seen from the rankit tables. If the sample size is even, the median does not receive a rankit at all because it lies between two of the measurements and will not appear in column 1. In this case the rankits in the rankit table are all used twice, beginning and ending at the top of the rankit column.

Occasionally the ranking will reveal two or more measurements sharing the same value. The measurements are then said to be **tied**. If this occurs then a third column will be needed because the rankits corresponding to the tied values need to be averaged. This occurs in our example where there are two measurements of 33.9 with corresponding rankits 0.335 and 0.516. The rankit average, $(0.335 + 0.516)/2 = 0.426$, is substituted into column 3 for these two measurements. Care must be taken when ties involve the median since this may entail averaging positive and negative rankits, though this does not occur here.

Finally, the rankits, corrected for ties (column 3), are plotted against the ranked data (column 1). A best fit line is then drawn with a transparent ruler, paying

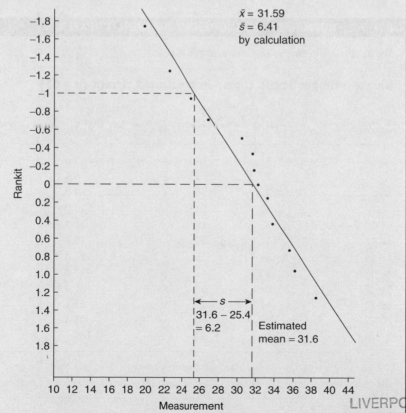

particular attention to the points close to the median. If a reasonably straight line is obtained the distribution may be taken as normal. If there is a *systematic* non-linear trend in the points the distribution is non-normal. In this example, the distribution is only approximately normal as it shows some convexity around the median.

If the data follow a normal distribution, an estimate of the sample mean is obtained by projecting the rankit value 0.00 to the line and then reading off the value on the abscissa. The sample standard deviation can be estimated in a similar way by projecting the +1 or −1 rankit to the line and finding the difference from the mean. In our example the mean is estimated as 31.6 and the standard deviation 6.2, fairly close to the values calculated from the raw data which are shown in the figure.

7.3 Graphical method using probability paper

A graphical method with probability paper can be used where the number of measurements exceeds 30, since rankit tables are not available for large samples. The data are prepared as a frequency distribution using a suitable number of classes and the cumulative frequency distribution is obtained. This is then plotted on special graph paper as shown in Box 7.2.

BOX 7.2: NORMALITY TESTING USING PROBABILITY PAPER

The data used in this example are shown below.

Nitrate–nitrogen levels (ppm) in a polluted stream ($n = 61$)

1 Class (ppm)	2 f	3 Cumulative f	4 Percentage cumulative f
0–2	1	1	1.6
2–4	2	3	4.9
4–6	3	6	9.8
6–8	5	11	18.0
8–10	8	19	31.1
10–12	11	30	49.2
12–14	8	38	62.3
14–16	9	47	77.0
16–18	6	53	86.9
18–20	4	57	93.4
20–22	3	60	98.4
22–24	1	61	100

Column 2 gives the frequencies of the observations and the third column gives the cumulative frequency obtained by adding all of the frequencies for that class and all classes before it. For example, the cumulative frequency for class 4–6 is 1 + 2 + 3 = 6. The last class at the base of the column will therefore contain the total number of measurements, which is 61.

The right-hand column contains the cumulative frequencies of column 3 as a percentage of the total number of measurements. Thus, in class 4–6, the cumulative frequency is 6 and the percentage cumulative frequency $(6/61) \times 100 = 9.84$. The percentage cumulative frequency is shown in the figure below. Notice the S-shaped form of the curve.

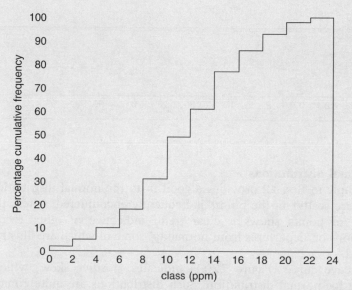

Finally, columns 1 and 4 are plotted on normal probability paper. The paper is oriented so that the probability scale runs from 0.01 to 99.9 along the x axis as shown in the following figure. The percentage cumulative frequency is plotted along this scale. You will find that the final value of 100 cannot be plotted, but this is not a problem since the important points lie near the median. The vertical axis, which is a linear scale, is used to plot the corresponding upper class limit. For example, in class 4–6, the value 6 is plotted on the ordinate against the percentage cumulative value of 9.8. Strictly, the class limits lie between 4 and 5.999, but rounding up the 5.999 to 6 is acceptable since graphing to more than three significant figures is impracticable.

For a normal distribution, the points will lie in a straight line, and if a reasonably straight line is evident a transparent ruler can be used to obtain an estimate of the sample mean and sample standard deviation. The line is drawn giving greatest weight to the points lying between cumulatives 25% and 75%.

The mean is estimated from the median, represented by the 50% cumulative. A line is drawn from the 50% value on the probability scale to the best fit line and the mean is estimated by projection (cf. rankits). The sample standard deviation is estimated likewise by projecting the 15.9% and 84.1% cumulatives on the probability scale, giving a standard deviation estimate either side of the mean.

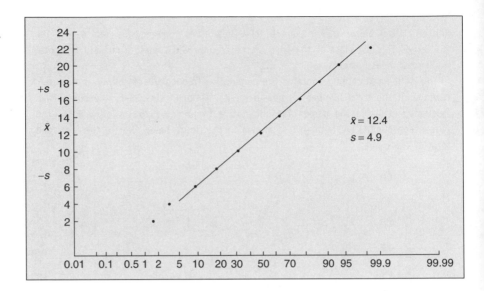

Non-normal distributions

The example in Box 7.2 provides a good fit to the normal distribution, but much more scatter in the points is frequently encountered. Often, the distribution of points shows a clear trend indicative of other recognised distributions or 'departures from normality', some of which are illustrated in Figure 7.2.

A concave curve (Figure 7.2(a)) indicates 'positive skew', which may suggest a log-normal distribution. Such distributions are quite common in nature, e.g. many animal and plant populations as sampled with a quadrat, particle size distributions and mortality rates. Probit (log × normal probability) paper can be purchased to test for the log-normal distribution, or the logarithm of the upper class limit can be plotted on the ordinate of normal probability paper. A convex curve (Figure 7.2(b)) is indicative of negative skew. This is less often observed, but some binomial distributions show it. An S-shaped curve suggests kurtosis. This is a normality departure which affects the shape of the frequency curve, while the mean, median and mode remain equal in value. A leptokurtic distribution (positive kurtosis, Figure 7.2(c)) has items bunched around the mean, giving a sharp peak. A platykurtic distribution (negative kurtosis, Figure 7.2(d)) is characterised by a broad summit which falls rapidly in the tails compared with the normal distribution. Bimodal distributions, which are occasionally encountered in toxicity testing (Chapter 14), can also produce a sigmoid probability plot. Multimodal distributions are often seen in animal samples where several age-classes are present, and these produce an undulating wave-like curve when plotted. Rankit plots can be interpreted in the same way as probability paper plots, provided the curves are viewed with the rankits arranged along the x axis.

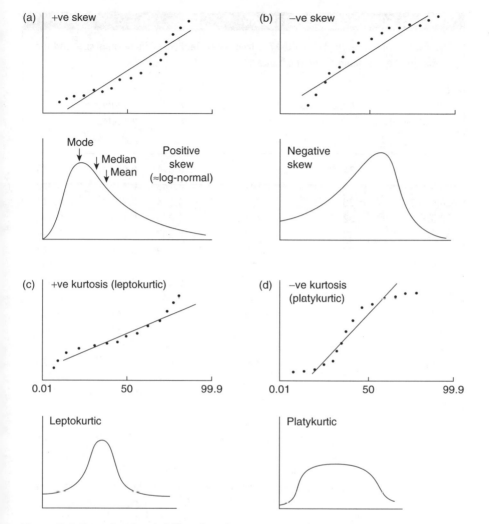

Figure 7.2 Example of probability plots for non-normal distributions and their interpretation: (a) positive skew (approximate log-normal distribution); (b) negative skew; (c) leptokurtosis; (d) platykurtosis.

7.4 Goodness-of-fit with the Kolmogorov–Smirnov one-sample test

The Kolmogorov–Smirnov method provides a test of goodness-of-fit to a specified distribution. Its use is explained in Box 7.3 using the data above for the rankit example.

The critical values of the Kolmogorov–Smirnov test statistic D are only

BOX 7.3: GOODNESS-OF-FIT TESTING

The data from the example in Box 7.1 have been ranked in increasing size and are shown in the first column of the table.

1 Ranked measurements x	2 Upper class limit U	3 Standardised deviate (U–x̄)/s	4 F	5 Cumulative expected frequency F̂	6 |d|
22.6	24	−1.569	0.071	0.058	0.013
25.0	26	−1.204	0.143	0.115	0.028
26.9	28	−0.838	0.214	0.200	0.014
30.5	32	−0.107	0.286	0.456	**0.170**
31.7	32	−0.107	0.357	0.456	0.099
31.9	32	−0.107	0.429	0.456	0.027
32.2	34	0.259	0.500	0.603	0.103
33.2	34	0.259	0.571	0.603	0.032
33.9	34	0.259	0.643	0.603	0.040
33.9	34	0.259	0.714	0.603	0.111
35.6	36	0.624	0.786	0.733	0.053
36.2	38	0.990	0.857	0.739	0.118
38.5	40	1.356	0.929	0.911	0.018
44.1	46	2.453	1.000	0.993	0.007

The numbers in the second column have been obtained from the frequency histogram, which was prepared with class intervals of two units. The column contains the upper class limits for each class in which the ranked x values belong. In the first row of column 2, the number 24 is the upper class limit (U) for the measurement of 22.6. Column 3 contains the standardised deviation of each measurement. To obtain this, the sample mean and standard deviation must be obtained (32.585 and 5.469, respectively). The standardised value in the first row is calculated as (24–32.585)/5.469. This is equivalent to the z score described in Chapter 5.

Column 4 shows the proportional cumulative frequency. The value in the first row is obtained by dividing the cumulative frequency by the total number of measurements, i.e. $1/14 = 0.0714$. As each row contains a single observation, each value down the table increases by 0.0714 to the last row of 1.00.

Column 5 gives values of the *cumulative expected frequencies* ($\hat{F}$) for the normal distribution. These can be obtained from z tables (**Table II**). The frequencies are found by obtaining the probabilities equivalent to the values obtained in column 3. For the negative values in column 3, the resulting probability is subtracted from 0.5 and entered into column 5. For positive values, the probability is added to 0.5 before being entered into column 5.

Finally, the modulus of the difference between the observed and expected frequencies in columns 4 and 5 is placed in column 6. The largest difference (0.170), shown in bold type, is the Kolmogorov–Smirnov test statistic D_{max}. This value is compared with critical values of D in **Table VII** for $n = 14$.

The critical value at $p = 0.05$ is **0.349**. As this is not exceeded by the sample statistic, there is no evidence to indicate that the data do not follow the normal distribution. A more sensitive form of this test is described by Harter *et al.* (1984), and an example of its use may be found in Sokal and Rohlf (1995).

published up to $n = 35$. For larger samples, D can be calculated as

$$D_{\text{crit}} = \sqrt{\frac{-\ln(\alpha/2)}{2n}}$$

where α is the desired significance level (usually 0.05).

Normality testing with Minitab
Place your data in column 1 of a Worksheet. The data do not need ranking; this will be done by Minitab.

Stat > basic statistics > normality test

In the dialog box, enter C1 in the 'Variable' box. Click on 'Anderson–Darling' and then click 'OK' to view the plot.

If the points lie approximately on a straight line, the distribution may be taken as normal. If you click 'SW test' and rerun the program, the p value generated can be used to reject the null hypothesis that the distribution is normal at the usual 0.05 level.

Deviations from normality can be interpreted in a similar way to graphical probability plots, but positive-skewed data will provide a convex (arched) curve, not a concave (valley) curve as found before.

Key notes

- The frequency distribution of a sample can often be identified with a theoretical distribution, such as the normal distribution.

- Four methods for comparing a sample distribution are: inspection of the frequency histogram; use of rankit tables; use of normal probability paper; the Kolmogorov–Smirnov one-sample test.

- Rankit plots can be used for small samples ($n < 30$) and provide estimates of the sample mean and standard deviation.

- Probability paper plots can be used for testing normal and log-normal distributions.

- Graphical methods often provide evidence of non-normal distributions, such as skewness and kurtosis.

- The Kolmogorov–Smirnov one-sample test provides a statistical test of an unknown distribution against a theoretical distribution.

Exercises

7.1 The following 29 measurements are of water temperature in degrees Celsius taken below a sewage outfall on consecutive days:

19.3 20.9 21.3 19.6 22.2 16.2 20.5 21.1 21.1 20.8 21.7
23.6 21.0 20.9 24.5 22.1 19.6 23.7 19.9 19.5 20.6 17.4
18.8 21.4 18.3 20.7 18.5 22.0 20.4

Plot the data as a frequency histogram using class intervals of 1 °C and calculate the mean. Now perform a test of normality using the rankit method and obtain an estimate of the mean and standard deviation.

7.2 The data below are of atmospheric ozone concentration measured in ppb from a monitoring station in a small seaside town ($n = 23$):

35.6 29.5 30.5 22.6 17.1 33.9 26.9 25.3 53.2 31.7 41.0
31.9 32.2 46.1 30.3 11.8 31.8 28.8 25.0 42.1 35.5 38.5
36.2

Test for a normal distribution using the rankit method, estimating the mean and standard deviation. Test your estimates against the calculated mean and standard deviation.

7.3 The measurements below are soil temperatures taken from a climate-change monitoring site in southern France. The data are plotted as frequencies for a total of 1139 measurements.

Soil temperature (°C)	Frequency f
14.10–14.199	0
14.20–14.299	4
14.30–14.399	10
−14.499	29
−14.599	52
−14.699	63
−14.799	93
−14.899	131
−14.999	115
−15.099	121
−15.199	103
−15.299	104
−15.399	81
−15.499	63
−15.599	58
−15.699	40
−15.799	39
−15.899	28
−15.999	0
−16.099	5

Plot the data as a frequency histogram indicating the mode. Obtain the percentage cumulative frequency and plot the measurements on normal probability paper. Estimate the sample mean and standard deviation.

7.4 The following data are of chemical oxygen demand in mg 1^{-1} taken from the outfall of a leather tannery plant $(n = 70)$:

8.32 5.01 15.19 16.08 10.53 7.88 13.53 3.91 18.68 10.7
14.04 9.31 14.58 13.08 11.55 8.03 6.91 12.98 10.67 15.02
16.88 21.24 5.91 1.66 10.32 14.51 9.08 7.22 17.49 13.01
11.51 18.51 23.27 10.41 14.38 13.62 14.58 16.35 21.13
2.27 6.85 10.37 8.52 13.36 20.93 8.68 6.51 19.05 10.13
5.36 17.47 15.38 5.44 12.4 14.01 18.76 8.68 6.59 6.20
17.13 12.59 15.02 10.67 9.32 16.08 13.33 8.80 11.12 15.59
10.60

Using class intervals of 2.0 mg 1^{-1} prepare the cumulative frequency distribution and test for a normal distribution using normal probability paper. Estimate the sample mean and standard deviation.

7.5 The following 20 measurements are of lead concentration (ppb) in children's teeth. Using class intervals of 1 ppb, plot the data as a histogram and perform a Kolmogorov–Smirnov one-sample test for a normal distribution.

7.3 10.2 9.8 15.5 6.1 5.8 7.8 8.3 10.3 11.8 16.2 7.9 4.3
7.1 8.8 11.3 7.4 5.7 12.4 9.9

CHAPTER EIGHT

Comparing two samples

Many situations arise where two sets of measurements of the same item need to be compared. For example, birds' eggshells are thought to be influenced by acid rain, which reduces the egg thickness index (mg cm^{-2}). Reduced egg shell thickness is thought to threaten the survival of some bird species. Suppose we investigate gulls' eggs at two nesting sites, taking a single egg at random from a number of nests at each site, so that the thickness index at the two sites can be compared. The yolk would be removed and the egg shell mass and surface area determined.

We may find that the measurements both of location (central value) and dispersion (spread or range) for the thickness indices differ at the two sites. If there is a difference only in the measure of location it is usually possible to employ a parametric statistical test based upon the difference between the two sample means. If the measure of dispersion also differs significantly

between the samples, we may still apply a parametric test, though subject to further procedures, or a non-parametric test could be used. All these methods are described below using a range of examples.

8.1 Some parametric methods – the two-sample z and t tests

If, in the above case, the thickness indices are known to belong to the same *population*, then the population means for the two sites (samples) will be the same and their difference will be zero. This difference between the means of two samples is the basis of the test.

The central limit theorem was introduced in Chapter 5 and it was noted that a collection of sample means will follow a normal distribution. It can also be shown that the distribution of the difference of two means $(\bar{x}_a - \bar{x}_b)$, where a and b are two samples, will also be normally distributed. Now recall the distribution of z, where simple statements of probability were made using normally distributed measurements and the standard normal curve. If the difference between two sample means can be divided by an appropriate standard deviation, a value of z can once again be calculated. The appropriate quantity is the *standard deviation (error) of the difference between two means* $(s_{\bar{x}_a - \bar{x}_b})$, which is obtained from the variance of samples a and b as follows

$$s_{\bar{x}_a - \bar{x}_b} = \sqrt{\left[\frac{1}{n}\left(s_a^2 + s_b^2\right)\right]}$$

When the difference between the two sample means is divided by the common sample standard deviation of a difference between two means, an estimate of z is obtained

$$z = \frac{\bar{x}_a - \bar{x}_b}{\sqrt{[(1/n)(s_a^2 + s_b^2)]}}$$

This equation can be used provided the number of measurements in each sample is reasonably large (>30, preferably >50), the measurements are approximately normally distributed and the number of measurements in each sample is the same.

To test for a difference between the population means, the null hypothesis is written H_0: $\mu_a = \mu_b$. This is equivalent to stating that the difference between the means is zero. The alternative hypothesis H_1 is $\mu_a \neq \mu_b$. With two samples whose means are similar, the difference between them will be small, as will the calculated value of z. Small values of z indicate acceptance of H_0, but critical values of z must be consulted to complete the test. These values are normally incorporated into t tables (**Table III**).

BOX 8.1: EXAMPLE OF A z TWO-SAMPLE TEST

The following data were obtained at two gull nesting sites. Fifty eggs were taken at random from each site, with one egg taken randomly from each nest. The relevant statistics are shown in the table.

Egg thickness index results (mg cm^{-2})

Site (sample)	a	b
n	50	50
$\bar{x}$	33.2	31.89
s^2	16.41	17.39

The difference between the two means is

$$\bar{x}_a - \bar{x}_b = 33.2 - 31.89 = 1.31$$

Note that it is immaterial which sample is labelled a or b. The difference between the means might just as well have been negative (i.e. -1.31). This does not affect the result for a two-tailed test but it could for a one-tailed test (see below).

The standard deviation of the difference is

$$\sqrt{\left[\frac{1}{n}\left(s_a^2 + s_b^2\right)\right]} = \sqrt{\left[\frac{1}{50}\left(16.41 + 17.39\right)\right]}$$

$$= \sqrt{(0.02 \times 33.8)}$$

$$= \sqrt{(0.676)}$$

The value of z can now be obtained as

$$\frac{\text{difference between the means}}{\text{standard deviation}} = \frac{1.31}{0.822}$$

$$= 1.594$$

Now turn to **Table III** and obtain the critical value for z (two-tailed) by going to the lowest row in the table marked with the infinity sign (∞). For $p = 0.05$, the critical value of $z = $ **1.96**. Since our test value of z does not exceed this figure, the null hypothesis is accepted.

8.2 Two-tailed and one-tailed tests

In Box 8.1 the terms two-tailed test and one-tailed test were mentioned and require some explanation. When discussing hypothesis testing in Chapter 5, the probability of obtaining a random observation of $z = 1.96$ was considered using the standard normal curve. Both tails of the curve contained unlikely values of z. In the example above, we found that the value of z was such that the null hypothesis H_0 was accepted. However, it would be useful for

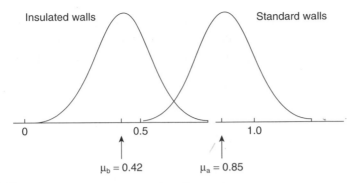

Figure 8.1 Two normal curves illustrating a one-tailed test.

us to look more closely at the situation if H_0 had been rejected. If we had accepted the alternative hypothesis $\mu_a \neq \mu_b$, what can we say about the two samples? If the two sample means do not come from the same normal population with a specified population mean, then they come from another population other than that specified by the null hypothesis. Since in general we shall not know which of the means is greater than the other we reject extreme values in either direction (i.e. in both tails); this procedure is known as a two-tailed test.

Occasionally sufficient information is available for us to specify the direction of the population means. For example, a study might be undertaken to measure the heat loss through brick walls built from two different types of brick. Suppose the sample mean for the standard brick wall was found to be 0.85 W m^{-1} K (the K stands for a temperature difference of 1 °C from one side of the wall to the other). This will be our estimate of the parametric mean (μ_a). If we now measure heat loss from a wall of insulation bricks, it is reasonable to expect that the heat loss will be less than that of the standard brick wall. We could therefore frame the alternative hypothesis as $\mu_a > \mu_b$. Figure 8.1 shows the two possible normal distributions under the new hypothesis.

The distribution to the right represents the null hypothesis with a parametric mean (μ_a) of 0.85 W m^{-1} K. The distribution to the left shows the hypothetical distribution for the insulated walls, whose mean will be less than that of the uncladded walls. We now calculate a value of z from samples a and b as shown in the z test above, where the estimate of μ_b is 0.42 W m^{-1} K with the standard deviation of the difference of the means 0.13 W m^{-1} K. We get

$$z = (0.85 - 0.42)/0.13$$

$$= 3.31$$

This is a large value of z, significant at $p = 0.01$. We therefore reject the null hypothesis and accept the alternative hypothesis that the two means are significantly different. This is what we should expect, comparing an insulating brick to a standard brick. Supposing, however, that the estimate of μ_b had been 1.28 W m^{-1} K. Then

$$z = (0.85 - 1.28)/0.13 = -3.31$$

In this situation, the alternative hypothesis is more unreasonable than the null hypothesis, as it is suggesting that the insulation brick conducts far more heat than the standard, and could not have been accepted. All positive values of z are more likely to occur on the null hypothesis than the alternative hypothesis and in this case we cannot reject values in the right-hand tail. The rejection region must all lie in the left-hand tail, so a test at the 5% level of significance must use a value of z which will exclude 5% of the left-hand tail. The corresponding value of z from **Table II** is **1.64**, and this is the value which must be used in the test. You will notice in many of the critical tables that values for both one- and two-tailed tests are indicated in the column headings. For most analyses, two-tailed tests are used, but it is important to think carefully about the hypotheses before conducting a test.

8.3 The difference between two sample means with limited data

If the number of measurements is less than 30 in each sample, the above method gives an unreliable estimate of z and another method is required. The problem was solved by 'Student', who introduced the t *test* early in the twentieth century. The test procedure is similar to the z test described above but, instead of referring to z, a value of t is required. Critical values of t for a specified probability are located in **Table III** using degrees of freedom which range from 1 to 30 in integer steps with a few larger values to infinity. The t distribution is similar to the normal distribution insomuch as the curve is symmetric about the median and is asymptotic. For all degrees of freedom below infinity the curve appears leptokurtic compared with the normal distribution and this property becomes extreme at small degrees of freedom (Figure 8.2). In practice, the table is used for degrees of freedom ranging from about five to 50. Above 50 degrees of freedom the z distribution (equivalent to t for infinite degrees of freedom) is used.

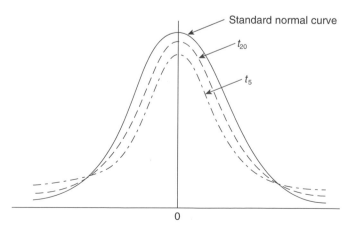

Figure 8.2 Two t curves for 5 and 20 degrees of freedom plotted with the normal curve.

BOX 8.2: A t TEST WITH EQUAL MEASUREMENTS IN EACH SAMPLE

In this example the chemical oxygen demand (COD) is measured at two industrial efflu-ent outfalls, a and b, as part of a consent procedure. The data are shown in the table. The calculations proceed as before, beginning with the difference between the means.

Chemical oxygen demand of effluent (mg l⁻¹)

	a	b	a	b
	3.48	3.89	3.20	3.55
	2.99	3.19	4.40	2.40
	3.32	2.80	3.85	2.99
	4.17	4.31	4.52	3.08
	3.78	3.42	3.09	3.31
	4.00	3.41	3.62	4.52
$\bar{x}$	3.701	3.406		
s^2	0.2571	0.3662		
$n = 12$ for both samples				

$$\bar{x}_a - \bar{x}_b = 3.406 - 3.701 = -0.292$$

The standard deviation of the difference between the means $(s_{\bar{x}_a} - s_{\bar{x}_b})$ is

$$\sqrt{\left[\tfrac{1}{n}\left(s_a^2 + s_b^2\right)\right]} = \sqrt{\left[\tfrac{1}{12}\left(0.3662 + 0.2571\right)\right]}$$

$$= \sqrt{(0.05194)} = 0.2279$$

$$t = \text{difference between means}/(s_{\bar{x}_a} - s_{\bar{x}_b})$$

$$= -0.292/0.2279 = -1.281$$

Now turn to the critical values of t (**Table III**) partly reproduced below. Note the way the tables are laid out with degrees of freedom (df) in the left-hand column and, above the remaining columns, six levels of significance ranging from 0.20 to 0.001 for a two-tailed test and from 0.10 to 0.0005 for a one-tailed test.

df	Level of significance for one-tailed test					
	.10	.05	.025	.01	.005	.0005
	Level of significance for two-tailed test					
	.20	.10	.05	.02	.01	.001
19	1.328	1.729	2.093	2.539	2.861	3.883
20	1.325	1.725	2.086	2.528	2.845	3.850
21	1.323	1.721	2.080	2.518	2.831	3.819
22	1.321	1.717	2.074	2.508	2.819	3.792
23	1.319	1.714	2.069	2.500	2.807	3.767
24	1.318	1.711	2.064	2.492	2.797	3.745
25	1.316	1.708	2.060	2.485	2.787	3.725

With a two-sample t test, there are $2n - 2$ degrees of freedom ($24 - 2 = 22$ in this case). Since the test is two-tailed – no prediction has been made regarding the direction of the test – the critical value ($p = 0.05$) is located in the $p = 0.05$ two-tail column with 22 df and is **2.074**. It is customary to indicate where our t value was taken from the tables and this can be done using two **subscripts**, one indicating the significance level and the other the degrees of freedom. In this case it is $t_{0.05, 22}$. Our negative test value is changed to positive (i.e. the modulus of the value is taken) as the test is two-tailed and it is seen to be less than the critical value. Thus the null hypothesis that the two population means are equal.

8.4 The two-sample t test for different numbers of measurements in each sample

The formula in Box 8.2 cannot be used when the numbers of measurements in each sample differ. In this case the standard deviation calculation in the **denominator** has to be modified, as shown in Box 8.3.

It is worth noting that there is an alternative formula for the standard deviation calculation when $n_a \neq n_b$, namely s.d. $= \sqrt{(s_a^2/n_a + s_b^2/n_b)}$. This may be used if the numbers of measurements exceed about 30, but the formula in Box 8.3 may equally well be used.

Occasions may arise when the degrees of freedom are such that the critical t value cannot be found in the tables. If df $= 36$, for example, you will not find any critical values for $t_{0.05, 36}$ in **Table III**. If a critical value is required it can be estimated using graphical interpolation (Figure 8.3). The critical t values above and below the required value are plotted along the ordinate against their degrees of freedom on the abscissa. The points are then joined by a straight line. The value of $t_{0.05, 36}$ can be obtained by locating 36 on the drawn line and finding the corresponding t value. The estimate will not be precise because the line joining the two points should be slightly curved, but it is sufficiently accurate for the t test. The same method can be applied to the other critical tables where critical values need interpolation.

BOX 8.3: A t TEST WITH DIFFERENT VALUES OF n IN THE SAMPLES

The data in the table show some more results from another two effluent sampling points. In this case the investigator dropped three of the bottles containing water from sample d, so $n_c \neq n_d$ and a pooled standard deviation will be used in the denominator. A t test is appropriate because both n_c and n_d are less than 30. In this case the sampling points c and d are situated above and below a reed bed, respectively. Since reed beds are known to remove COD from effluents it is hypothesised that $\mu_{above} > \mu_{below}$, while the null hypothesis is given as $\mu_{above} \leq \mu_{below}$. This is an example of a one-tailed test.

COD analyses for two sites (mg l^{-1})

	Sample c (above reed bed)	Sample d (below reed bed)
	3.53	2.98
	3.81	3.42
	2.88	3.18
	3.32	2.80
	4.14	2.60
	3.04	3.61
	3.45	3.09
	3.75	3.27
	4.29	2.36
	3.14	3.82
	3.22	2.88
	2.61	3.17
	3.10	
	3.68	
	4.52	
n	15	12
$\bar{x}$	3.499	3.098
s^2	0.2898	0.1701

For a one-tailed test the direction of the difference between the means has been specified as $\mu_c - \mu_d$. In the calculation this is approximated by $\bar{x}_c - \bar{x}_d$ and it is important to calculate this difference and *not* $\bar{x}_d - \bar{x}_c$. For our example, $\bar{x}_c - \bar{x}_d$ is $3.499 - 3.098 = 0.401$, a positive number. If the difference had been negative the null hypothesis would have been accepted as the one-tailed test specified, $\mu_c > \mu_d$.

The pooled standard deviation of the difference between the means is given by a rather long formula when $n_c \neq n_d$

$$t = (\bar{x}_c - \bar{x}_d)\left\{\sqrt{\left[\frac{(n_c - 1)s_c^2 + (n_d - 1)s_d^2}{n_c + n_d - 2}\frac{n_c + n_d}{n_c n_d}\right]}\right\}^{-1}$$

This gives for the denominator

$$\sqrt{\left[\frac{(15 - 1) \times 0.2898 + (12 - 1) \times 0.1701}{15 + 12 - 2}\frac{15 + 12}{15 \times 12}\right]}$$

$$= \sqrt{\left(\frac{4.0572 + 1.8711}{25}\frac{27}{180}\right)}$$

$$= \sqrt{(0.2371 \times 0.15)}$$

$$= 0.1886$$

$$t = \text{difference between means}/(s_{\bar{x}_c} - s_{\bar{x}_d})$$

$$= 0.401/0.1886 = 2.126$$

Now turn to the t tables and locate the row corresponding to degrees of freedom $n_a + n_b - 2$ $(15 + 12 - 2 = 25$ df). For $p = 0.05$ the critical value of t *one-tailed* (**Table III**) is **1.708**. This value is exceeded by the test value so the null hypothesis is rejected and we can state that the population means of the samples differ significantly in the specified direction. The reed bed does appear to be removing COD from the effluent.

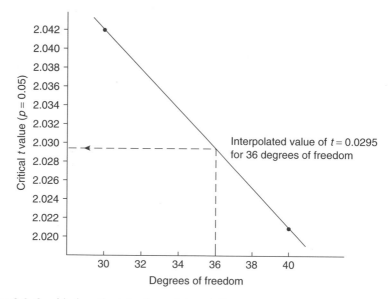

Figure 8.3 Graphical method for linear interpolation.

8.5 t and z test assumptions

For the t and z tests to be valid, the measurements should be at least approximately normally distributed and **homoscedastic** (with population variances equal). An approximate normal distribution can be checked using a stem and leaf plot, a normal quantile plot or another graphical method (Chapter 7). For most purposes, a stem and leaf plot should suffice. To check for homoscedasticity, divide the largest sample variance by the smallest to produce a variance ratio. If the ratio falls within the stippled area in Figure 8.4 the variances may be considered homoscedastic. For the example in Box 8.3, the two sample variances were 0.2898 and 0.1701, giving a ratio of $0.2898/0.1701 = 1.703$. Since in this case $n_a \neq n_b$ the smaller n ($n = 12$) is used, giving degrees of freedom $n - 1 = 11$ on the abscissa of Figure 8.4. The ratio can be seen to lie in the homoscedastic region. If the variances had been

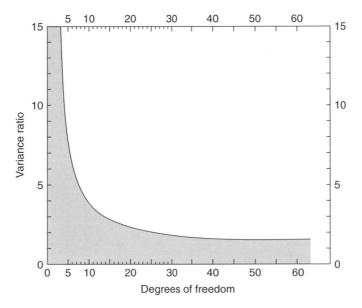

Figure 8.4 Graph for testing homoscedasticity when comparing two samples.

found to be **heteroscedastic** (i.e. population variances unequal, located in the clear area of Figure 8.4) then a log transformation may be attempted and the variance ratio replotted on Figure 8.4, or a Mann–Whitney test can be used.

Another important assumption for the *t* and *z* tests is that the samples must be independent, i.e. the values in one sample must have no influence over those in the other. Where this breaks down, a 'matched' or 'paired' test can be used.

Two-sample *t* tests using Minitab

Input the two sets of data into columns C1 and C2 as shown below. Note in this case that $n_a \neq n_b$, so an asterisk is inserted for the 'missing data'.

C1	C2
7.10	4.95
5.40	4.55
6.45	4.10
7.45	6.05
6.75	3.25
7.45	1.95
8.00	3.25
9.00	*

On the menu bar:

Stats > Basic statistics > 2-sample t

Click the box labelled 'Samples in different columns' and type in the boxes C1 and C2.

In the 'Alternative' box, ensure that 'not equal' is showing for a two-tailed test. This contains a statement of the alternative hypothesis H_1.

Click the 'Assume equal variances' box. Now click 'OK' and you should receive the output shown below. Note that Minitab also provides an estimate of the confidence interval for the difference between two means. It is obtained from the standard deviation of the difference in the same way as the confidence interval for the mean (Chapter 9).

```
Two sample T for C1 vs C2
              N           Mean            StDev           SE Mean
C1            8           7.20            1.07            0.38
C2            7           4.01            1.34            0.51
95% CI for mu C1 – mu C2: (1.84, 4.53)
T-Test mu C1 = mu C2 (vs not =): T = 5.12 P = 0.0002 DF = 13
Both use Pooled StDev = 1.20
```

You may also use the same box to perform a one-tailed test. Suppose there is good reason to assume that the population mean of the sample in C1 is greater than that in C2. In this case, change the text in the 'Alternative' box to 'greater than'. You may also wish to see what happens if 'less than' is entered in this box.

If you fail to click the 'Assume variances equal' box, a test is performed using reduced degrees of freedom, a method sometimes used to account for the effect of heteroscedasticity in the t test.

8.6 Non-parametric tests for two samples – the Mann–Whitney test

The Mann–Whitney test is a useful test for two samples, which can be used as an alternative to the t test, particularly where the assumptions for that test cannot be demonstrated. It can also be used for measurements made at ordinal level. It is usually assumed that the distribution of the measurements in the two samples are of the same general form (prepare a stem and leaf plot).

Non-parametric tests are also called 'distribution-free' tests because they are independent of the underlying population distribution, the only assumptions being independence of observations and continuity of the variable which is being measured. They can be used with a wider range of data than parametric tests and they are simpler to use. Most non-parametric tests involve a ranking procedure.

The test can be undertaken with any sample size. However, the method of calculating the test statistic (U) depends upon the size of n. Examples of these methods are detailed below. In the Mann–Whitney test, the null hypothesis cannot be stated in terms of population parameters, but is defined as the equality of the medians of the populations from which the two samples are drawn.

The method used when neither n_1 nor n_2 is greater than eight is demonstrated in Box 8.4. In this case, the observations of the two groups are combined and then ranked in increasing size.

When the number of measurements in the larger sample (n_2) is greater than eight, other methods of calculating U are used. This is the classical Mann–Whitney procedure described in Box 8.5.

BOX 8.4: MANN–WHITNEY TEST WHERE n IS SMALL

Suitable data are provided in the table.

Sample A	Sample B
9	10
12	15
6	11
7	13
	18

These measurements are combined and ranked in increasing size. Appended to each measurement is the sample from which it was taken (A or B).

A	A	A	B	B	A	B	B	B
6	7	9	10	11	12	13	15	18

Now count the number of A scores which precede a B score. The first A (corresponding to the value of six, far left) comes before five Bs. The second A also comes before five Bs. Adding up all the A scores which precede a B score gives a total of $5 + 5 + 5 + 3 = 18$. Now count the number of B scores which precede the A scores. This is $1 + 1 = 2$. The smaller of the sums is the Mann–Whitney test statistic (U) which has the value 2 in this case. This test value may now be compared with tabulated values of U, such as those given by Siegel (1956). The probability of occurrence under H_0 (two-tailed) is 0.032, so the null hypothesis of equality of the medians would be rejected.

BOX 8.5: MANN–WHITNEY TEST WHERE n_2 LIES BETWEEN 9 AND 20

Soil moisture content has a profound influence upon the use to which that soil is put. Soils with a high moisture content tend to be put under permanent pasture in the agrarian regions of Europe, while those with a lower moisture content are often planted in rotation with arable crops. The following data provide representative soil moisture contents on north- and south-facing slopes under grassland in June. The Mann–Whitney test will be used to test the null hypothesis that the population medians of the two samples are the same. A variance ratio test will show that the measurements are heteroscedastic, failing one of the assumptions of the t test. Also, the measurements are percentages, which would not be expected to follow a normal distribution. In this case the measurements will be ranked, but a formula is used to generate the U statistic.

Soil moisture content (% dry wt)

North-facing	Rank	South-facing	Rank
70.1	27	50.5	12
66.1	19	49.4	11
66.5	20	41.6	6
69.4	26	33.4	2
71.3	29	45.3	7
67.6	22	55.9	15
68.0	23	60.0	16
67.2	21	38.3	3
73.0	31	39.5	4
72.2	30	40.6	5
63.6	17	70.7	28
65.4	18	46.4	8
69.1	24.5	53.0	13
69.1	24.5	26.7	1
		47.1	9
		54.2	14
		48.5	10
$n_1 = 14$	$\sum R_1 = 332$	$n_2 = 17$	$\sum R_2 = 164$

The measurements are first combined and ranked from lowest to highest and the ranks are placed next to the measurements as shown above. In this example, one tie occurred. The quantities n_1 and $\sum R_1$ (sum of ranks) are assigned to the smaller n when the sample sizes differ. Ranking by hand is particularly error-prone and is made easier by writing out the total number of digits in the combined data (in the above case $14 + 17 = 31$) in a line. An alternative is to use a stem and leaf plot.

tie

$$\cancel{1} \cancel{2} \cancel{3} \cancel{4} \cancel{5} \dots \cancel{22} \cancel{23} \cancel{24} \cancel{25} \cancel{26} \cancel{27} \ 28 \ 29 \ 30 \ 31$$

The digits are crossed off as the ranks are assigned, which provides a check on the tally as the ranks accumulate.

Two values of U are obtained as

$$U_1 = n_1 n_2 + [n_1(n_1 + 1)]/2 - \sum R_1$$

$$U_2 = n_1 n_2 + [n_2(n_2 + 1)]/2 - \sum R_2$$

giving

$$U_1 = 14 \times 17 + 14(15)/2 - 332$$

$$= 238 + 105 - 332 = 11$$

$$U_2 = 14 \times 17 + 17(18)/2 - 164$$

$$= 238 + 153 - 164 = 227$$

A useful check on these values of U can be obtained using the relationship

$$U = n_1n_2 - U'$$

i.e.

$$227 = 238 - 11$$

It can also be used to obtain the required value of U using a single calculation, for U_1 say.

The test statistic is the smaller of the two values of U, namely 11. This value is compared with critical values using **Table IV**. The critical value is located with the values of n_1 and n_2. For a two-tailed test and $p = 0.05$, the value from the tables is **67**. For this test, the null hypothesis is *rejected* if the test value is less than the critical value, which is the case here. Thus the medians of the two samples are significantly different.

8.7 Mann–Whitney test for large samples, $n_2 > 20$

Critical tables for U become unwieldy as n increases above 20 and Mann & Whitney obtained z as a function of U and n as shown below

$$z = \frac{U - n_1n_2/2}{\sqrt{[n_1n_2(n_1 + n_2 + 1)/12]}}$$

Suppose that the test value of U was found to be 148, with $n_1 = 16$ and $n_2 = 29$. Then

$$z = \frac{148 - (16 \times 29)/2}{\sqrt{[16 \times 29 \times (16 + 29 + 1)/12]}}$$

$$= (148 - 232)/\sqrt{(1778.7)}$$

$$= -84/42.174 = -1.99$$

For a two-tailed test, the modulus (+1.99) is taken.

With reference to **Table III** ($t_\infty = z$) the critical value of z is **1.96** at $p = 0.05$, which shows that the result is just significant for a two-tailed test.

A special case occurs where ties exist between the two samples. A correction term can be applied (e.g. Siegel, 1956) though the effect of ties is negligible unless there are large numbers of tied observations (where more than about 30% of the measurements are tied across the samples). Another useful non-parametric method for large samples is the Kolmogorov–Smirnov two-sample test.

Mann–Whitney test using Minitab

Input the measurements for the two samples in two separate columns, C1 and C2. In the dialog box

Statistics – Non parametrics – Mann-Whitney

In the boxes type C1 for 'First sample' and C2 for 'Second sample'. In the 'Alternative' box select 'not equal to' for a two-tailed test (a one-tailed test may also be performed). Click 'OK'.

A sample output is shown below. Note that an approximate 95% confidence interval is output as ETA1 – ETA2. The printout shows that the test is significant at $p = 0.2888$. In other words, the null hypothesis is accepted in this case since $p > 0.05$. The value W is the sum of the ranks in the first sample; the value of U is not given. ETA denotes the population median.

```
Mann-Whitney Confidence Interval and Test
C1    N = 10    Median = 12.700
C2    N = 10    Median = 14.200
Point estimate for ETA1-ETA2 is -1.000
95.5 Percent CI for ETA1-ETA2 is (-2.900, 1.100)
W = 90.5
Test of ETA1 = ETA2 vs ETA1 not = ETA2 is significant at 0.2899
The test is significant at 0.2888 (adjusted for ties)
Cannot reject at alpha = 0.05
```

8.8 Paired tests

A researcher wishes to compare the concentrations of dissolved phosphate in two stretches of the same river using the molybdenum blue method. At midday, a sample is taken upstream above a town, and the researcher then drives below the town to take a second sample downflow a few minutes later. The same procedure is followed for the next nine days to obtain a total of ten upstream and ten downstream samples, which are then analysed. Most of the soluble phosphate in polluted river systems is known to originate from discrete outfalls and the study was undertaken to see whether additional phosphate was discharged into the river during its course through the town. Framed as a hypothesis, the aim is to test whether point discharges in the town significantly increase the mean phosphate concentration below the town (Table 8.1).

Before undertaking a t test, it is necessary to check whether these samples were taken independently. When the investigation began on day 1 the weather was fine and phosphate increased downstream. This continued on the next day and the day after. Then the weather changed and heavy rain fell on days 4 and 5. The rain entered the river, increasing the discharge. If you look at the phosphate levels on days 4–6 you can see that,

Table 8.1 Soluble reactive phosphate concentration in river water taken above and below a town

| | Reactive phosphate (ppm) | | |
Day	Above town	Below town	Difference
1	0.294	0.475	−0.181
2	0.246	0.463	−0.217
3	0.277	0.374	−0.097
4	0.195	0.216	−0.021
5	0.135	0.194	−0.059
6	0.195	0.185	+0.010
7	0.240	0.290	−0.050
8	0.273	0.321	−0.048
9	0.244	0.359	−0.115
10	0.255	0.366	−0.111

initially, the overall level of phosphate in the river has fallen at both sampling points and that the difference in phosphate concentration between the sampling sites is reduced, with an actual fall in phosphate between stations on day 6. The differences must be related to the discharge. Since the same river is being sampled twice, effects such as rainfall must influence *both* sites, and the samples are not independent. The condition of the river above the town must influence its condition below the town. In such situations another form of the *t* test must be employed. This is known as the 'paired *t* test' and the two samples taken each day are taken as pairs. The null hypothesis is based upon a population mean difference of zero, about which the differences should be normally distributed. The condition of equal variance (homoscedasticity) still applies to the two samples and their difference should be normally distributed. Stem and leaf plots should be drawn for both samples. They will contain both *random* and *between pairs* variation. The failure to allow for differences between individual samples, by carrying out a normal *t* test for example, can lead to erroneous conclusions.

The paired *t* test using Minitab
Enter your data in two columns of a worksheet, C1 and C2. The column length must be the same for both samples. The test is undertaken on a single column which must contain the differences between the pairs.
    On the menu bar:

Stat > Basic statistics > paired t

In the dialog box, type C1 in 'First sample' and C2 in 'Second sample', noting that the difference will be calculated as C1 − C2. Now click 'OK'.
    A typical output is shown below. Note that the confidence interval for $\bar{D}$ is also given.

BOX 8.6: PAIRED t TEST CALCULATION

For a paired t test, the analysis is conducted on the column of *differences* between the pair values shown in Table 8.1. The mean of the differences, usually written $\bar{D}$, is then obtained, care being taken to add the negatives to the positives (i.e. $-0.059 + 0.01 = -0.049$), and the sum is divided by the number of pairs, which is ten in this case. In our example $\bar{D} = 0.889/10 = 0.0889$.

The sample standard deviation of the differences is found in the usual way

$$s_D = \sqrt{\frac{0.123\ 29 - (0.886)^2/10}{10 - 1}} = 0.070\ 39$$

From this quantity the standard error of the mean is obtained as

$$s_{\bar{D}} = s_D/\sqrt{n}$$

$$= 0.070\ 39/3.1623 = 0.022\ 26$$

To obtain a test value of t, the mean of the differences is divided by its standard error

$$t = \bar{D}/s_{\bar{D}}$$

$$= 0.0889/0.022\ 26 = 3.994$$

Now consult the t tables with $n - 1$ degrees of freedom (*not* $2n - 2$ as was the case with the independent t test). The test is one-tailed because the null hypothesis H_0 states that the phosphate concentration below the town will not be greater than that above the town (i.e. $\mu_{below} \leq \mu_{above}$). The alternative hypothesis H_1 states the mean phosphate concentration, $\mu_{below} > \mu_{above}$. The critical value of t (**Table III**) (one-tailed, $p = 0.05$) is **1.833**. The test value of t exceeds this figure, so the null hypothesis is rejected. In other words, the phosphate concentration increases significantly below the town.

```
Paired T for C1 - C2

                    N         Mean        StDev       SE Mean
      C1           10        13.230       1.796        0.568
      C2           10        14.170       2.129        0.673
Difference         10        -0.940       1.157        0.366
95% CI for mean difference: (-1.768, -0.112)
T-Test of mean difference = 0 (vs not = 0): T-Value = -2.57
P-Value = 0.030
```

If a one-tailed test is required, change the settings in the 'Alternative' box as shown under the ordinary t test above. It is important to ensure that the populations from which the samples are drawn are placed in the correct order when the difference between C1 and C2 is obtained. If the number of observations in each sample exceeds 30 use *1 sample z*.

8.9 A non-parametric paired test – Wilcoxon's matched pairs signed ranks test

An alternative method for paired measurements is based upon a ranking procedure. Wilcoxon's test is also used when the assumptions for the paired t test fail as, for example, when the measurements are not normally distributed. A test statistic T is obtained by a ranking procedure and is compared with critical values of T. In common with the non-parametric Mann–Whitney test for two independent samples, the null hypothesis is *rejected* if the test value is found to be less than the critical value found in the tables. For parametric tests, and many other non-parametric tests, you will recall that the null hypothesis for a two-tailed test is accepted under these conditions.

BOX 8.7: WILCOXON MATCHED PAIRS SIGNED RANKS TEST

The data below are used to illustrate the test procedure. A long-term experiment is conducted to determine if total soil nitrogen is depleted in grassland which is given one cut per year but is left unfertilised. The study will show whether atmospheric sources of nitrogen are sufficient to replenish the nitrogen lost by cutting, denitrification and leaching. Twelve 1 m² grassland plots were maintained for a period of ten years, and from each plot the total nitrogen was measured before and after the experiment. The 12 measurements may be considered paired as they are taken from the same 12 sites over the ten year period. The hypothesis under test is that nitrogen content before and after the ten year period has not changed.

Total soil nitrogen (kg m⁻²)

1 Initial N content	2 N content after 10 yr	3 Difference	4 Difference minus sign	5 Rank	6 Signed rank
0.347	0.310	0.037	0.037	9	9
0.373	0.323	0.050	0.050	12	12
0.346	0.327	0.019	0.019	5	5
0.349	0.356	−0.007	0.007	1	−1
0.326	0.337	−0.011	0.011	3	−3
0.370	0.333	0.037	0.037	10	10
0.355	0.345	0.010	0.010	2	2
0.360	0.321	0.039	0.039	11	11
0.378	0.350	0.028	0.028	6.5	6.5
0.367	0.353	0.014	0.014	4	4
0.409	0.374	0.035	0.035	8	8
0.332	0.360	−0.028	0.028	6.5	−6.5

The results of the experiment are laid out in columns 1 and 2. In column 3 the differences between the treatments are given. They may be either positive or negative and in this case most of the differences are positive. If there had been any zero values in this column because the 'before' and 'after' observations were the same, they would be excluded from the following procedure. In column 4 the negative signs have been removed from the differences prior to the ranking. Ranking of the 'unsigned' differences is shown in the next column. Note the single tie. Finally, in column 6 any negative signs of column 3 are reassigned to the ranks. Column 6 is the 'working column' used for calculating the test statistic. In this example only three of the differences are negative, giving a total of three negative ranks in column 6.

The test statistic T is obtained by summing the positive and negative ranks separately. The smaller of the sums is assigned T.

$$\text{Sum of positive ranks: } 9 + 12 + 5 + 10 + 2 + 11 + 6.5 + 4 + 8 = 67.5$$

$$\text{Sum of negative ranks: } 1 + 3 + 6.5 = 10.5$$

$$\text{Test statistic } T = 10.5$$

This value is now compared with the appropriate critical value (**Table V**). For a two-tailed test with $n = 12$ pairs and $p = 0.05$ the critical value is **14**. The critical value exceeds the test value and the null hypothesis is rejected, and the alternative hypothesis is accepted.

Wilcoxon's test with Minitab

Enter the data into columns C1 and C2 as pairs.

Stat > basic statistics > nonparametrics > 1-sample Wilcoxon

In the dialog box, enter C1 C2 in the 'Variables' box and then click on 'Test median'. For a two-tailed test, ensure that 'not equal' is showing in the 'Alternative' box; now click 'OK'.

Key notes

- Comparisons between two samples can be made with reference to the difference between the sample means or medians.

- Comparisons between sample means are made by calculating their difference and dividing by an appropriate sample standard deviation.

- Where the sample size is large ($n > 50$) a two-sample z test can be used. For smaller samples, a t test can be used.

- The parametric t and z tests should be applied to independent interval/ratio measurements which are at least approximately normally distributed.

- A simple test of homoscedasticity should be applied prior to application of t and z tests.

- For non-normally distributed or heteroscedastic data, the non-parametric Mann–Whitney test can be used.

- Where the measurements obtained in one sample can be shown to depend on measurements in the other sample, paired tests should be used.

- A paired t or z test can be used for dependent interval/ratio data which are at least approximately normally distributed.

- The non-parametric Wilcoxon matched pairs signed ranks test can be used for paired interval/ratio data sets.

Exercises

8.1 Zinc concentrations were measured on 12 random samples taken from two industrial waste lines, E and F, and the results in 10^{-8} M Zn are given below. Test for a significant difference between the means using t.

Site E: 12.5 12.1 13.4 13.0 12.1 12.15 12.25 12.7 11.4
11.7 12.5 13.7

Site F: 9.8 11.6 10.4 10.3 10.4 11.1 9.8 10.15 10.2 10.8
10.9 11.2

8.2 An investigation of the effects of two different atmospheric carbon dioxide concentrations on the growth of a weed *Senecio jacobaea* was undertaken under controlled conditions. Eighteen plants from the same seed stock were grown from seed to maturity in 12 weeks under CO_2 concentrations of 370 and 450 ppm. Each plant was then harvested, dried and weighed in grammes. Perform a t test to determine if the 'high CO_2' plants (450 ppm) produced more biomass than the control (370 ppm) plants.

CO_2, 370 ppm (control): 22.9 19.3 17.0 20.2 21.8 19.8 23.6
24.1 25.3 24.7 23.1 23.9 25.2 18.1 17.6 22.5 23.0 23.8

CO_2, 450 ppm: 25.8 25.4 22.3 26.1 26.0 19.9 18.7 21.3
22.7 26.4 27.1 17.7 20.8 24.6 23.9 25.9 26.0 25.5

8.3 Concentrations of lead (Pb in ppb) were measured in the plasma of a random sample of 8-year-old boys from a UK city and another sample of 8-year-old boys from the surrounding countryside in 1968. It was hypothesised that the city boys would have elevated levels of lead compared with the country boys. Perform a suitable t test of the two samples.

City boys: 5.36 6.48 7.19 5.55 4.89 4.12 6.00 6.65 6.80
5.00 3.50 5.72 6.15 6.30 ($n = 14$)

Country boys: 4.40 5.25 4.50 4.80 5.49 2.95 3.25 3.93
4.95 5.65 4.15 3.41 3.68 ($n = 13$)

A further study was made of 14 8-year-old girls from the same city. The girls had a mean plasma Pb level of 5.51 ppb with sample standard deviation 0.9318 ppb. How did they compare with the city boys?

8.4 A comparison was made between the total bacteria counts obtained from swabs taken from two designs of telephone mouthpiece, A and B. Both mouthpieces were subjected to two months' use by randomly selected volunteers so that all headpieces received the same amount of use. Use a Mann–Whitney test to determine whether the level of bacterial contamination ascertained by the swabs differed between the two designs.

Total bacterial counts $\times 10^5$

Design A: 6.8 3.2 10.9 7.8 4.1 12.1 5.7 1.1 3.7 8.4 9.6
10.6 8.9 5.4 2.0

Design B: 8.7 4.4 15.1 2.9 6.8 14.8 11.3 6.3 10.0 1.1
13.1 14.3 9.4 13.7 15.6

8.5 Houses were surveyed for rising damp using electrical resistance (relative units) in the north and south side of an east–west running street. Use a Mann–Whitney test to determine if there is a significant difference in rising damp between the two sides.

North side: 1842 1041 495 842 1541 1980 222 395 1615
2042 930 898 572 1315 741 1251 1092 1695
1225 1138 1161 132 675 ($n = 23$)

South side: 1265 563 868 449 1205 1690 1855 1372 672
1131 1508 1770 1632 2031 1088 1444 721
610 756 1180 1399 ($n = 21$)

8.6 The following temperatures (°C) were taken 10 m above and 10 m below a sewage outfall on a stream. Temperatures were taken at a fixed time on 15 occasions. Use a paired t test to determine if there is a significant increase in temperature below the outfall.

Sampling no.	Above outfall	Below outfall
1	6.8	7.9
2	5.1	6.4
3	7.2	8.8
4	8.8	9.9
5	9.2	10.6
6	9.5	9.1
7	10.6	9.8
8	12.2	13.7
9	14.6	14.0
10	15.8	15.2
11	14.5	13.9
12	13.1	14.8
13	10.6	13.1
14	8.4	7.9
15	9.2	11.5

8.7 As a quality control measure, 14 ammonia-sensing electrodes were each dipped into two solutions containing the same concentration of ammonia but different concentrations of salt (sodium chloride). The data below show the level of ammonia detected by the electrodes (in ppm) without salt (A) and with 1 g l^{-1} salt (B). Perform a paired t test to determine if the presence of salt affects the readings.

Electrode no.	A	B
1	4.363	4.360
2	4.428	4.492
3	4.380	4.378
4	4.409	4.400
5	4.306	4.308
6	4.319	4.312
7	4.336	4.256
8	4.408	4.410
9	4.388	4.360
10	4.291	4.231
11	4.388	4.326
12	4.408	4.406
13	4.351	4.388
14	4.367	4.361

8.8 Fourteen health workers were randomly selected to test the efficacy of a germicidal handcream. Total viable bacterial counts were determined by pressing the fingertips of the nurses onto nutrient agar plates before an 8 h shift. The cream was applied and the procedure was repeated at the end of the shift. Use the Wilcoxon matched pairs signed ranks test to determine if the cream significantly reduced bacterial counts.

	Total viable counts $\times 10^3$	
Nurse	Before	After
1	3.61	1.04
2	2.82	0.41
3	1.98	1.71
4	4.21	1.31
5	0.82	1.00
6	0.31	0.42
7	2.25	1.33
8	6.41	0.68
9	1.23	0.91
10	2.07	0.22
11	1.02	1.15
12	0.98	0.31
13	3.85	0.16
14	2.33	1.18

CHAPTER NINE

Confidence intervals for means and proportions

Means are among the most frequently used descriptive statistics and a researcher often needs to know how close a sample mean is to the population mean (μ). For example, there have been several large-scale studies of the amount of dissolved/dispersed petroleum hydrocarbons (DDPH) in the upper ocean. It would be impossible to sample the entire ocean so the population mean at any time is unknown. However, a sample mean can be obtained. It has already been mentioned that the sample standard deviation (or *standard error*) of this mean will be normally distributed and obtained as

$$s_{\bar{x}} = s/\sqrt{n}$$

This tells us that the population mean lies within one standard deviation of the sample mean with about 68% confidence. It is usual to quote 95%

BOX 9.1: CONFIDENCE INTERVAL CALCULATION FOR A LARGE SAMPLE ($n > 30$)

Fifty analyses of DDPH were made at stations scattered throughout the Atlantic Ocean. The sample mean was found to be 4.75 µg l^{-1} with sample standard deviation 3.99 µg l^{-1}. Find the 95% confidence intervals to the mean.

$$\text{standard deviation of the mean} = s/\sqrt{n} = 3.99/\,7.071 = 0.5643$$

95% confidence limits are obtained by multiplying this value by $z_{0.05}$ or **1.96**, i.e. $0.5643 \times \textbf{1.96} = 1.105$. We may then state that the true (population) mean lies between the values of $\bar{x} + 1.105$ and $\bar{x} - 1.105$ with 95% confidence, i.e. between the concentrations $4.75 + 1.105$ and $4.75 - 1.105$ (5.855 and 3.644 µg l^{-1}). Had the sample been larger, say $n = 100$, the confidence limits would have been considerably narrower since the sample standard deviation would have been divided by ten rather than 7.071. Doubling n, and assuming that the sample standard deviation remains unchanged, reduces the confidence interval by about 70%.

BOX 9.2: CONFIDENCE INTERVAL CALCULATIONS FOR SMALL SAMPLES ($n < 30$)

Ten measurements of DDPH were obtained from a stretch of coast along the Indian Ocean and yielded a mean of 78.2 µg l^{-1} with sample standard deviation 38.6 µg l^{-1}. Find 95% confidence intervals.

For small samples, the 95% confidence intervals are obtained as

$$\bar{x} \pm t_{0.05,\ n-1}s/\sqrt{n} = 78.2 \pm \mathbf{2.262} \times 38.6/\sqrt{(10)}$$

$$= 78.2 \pm 27.61 \text{ µg } l^{-1}$$

The appropriate value of t is found with $n - 1$ degrees of freedom (i.e. $10 - 1 = 9$ df) under $p = 0.05$ in **Table III**, giving a value of **2.262** (two-tailed). Multiplication by this value of t increases the confidence interval, as t is always numerically greater than z for a given probability level.

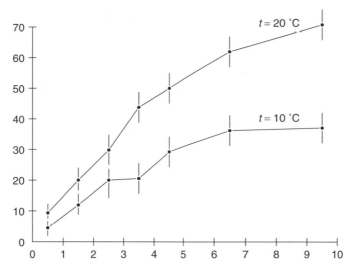

Figure 9.1 95% confidence intervals attached to two line graphs plotted on common axes.

confidence intervals by multiplying the sample standard deviation of the mean by the appropriate value of z, namely **1.96**.

Provided the number of measurements is above 30, a value of z is used to generate confidence intervals. Intervals may also be calculated with other values of z, i.e. multiplication of the standard error of the mean by **1.645** yields 90% intervals and by **2.576** yields 99% intervals.

If the sample size is less than 30, the standard deviation of the mean must be multiplied by an appropriate value of t to generate confidence intervals. This is because the sample standard deviation (s) only provides a good estimate of the population standard deviation (σ) when n exceeds 30.

It is common practice when graphing data such as time series to show the mean and 95% confidence interval (Figure 9.1). Often, the mean ±2 standard errors is plotted in published work. When n is large for each sample, the appropriate z value for 95% limits (**1.96**) is rounded to the value 2.

Confidence intervals with Minitab
Confidence intervals for the mean are obtained from the one-sample t test as follows:

Stat > basic statistics > 1-sample t

In the 'Variables' box enter the columns containing the samples with each sample contained in a separate column. Click 'Confidence interval'. The default value is the usual 95%. Now click 'OK' to output the intervals. The intervals can be displayed graphically if desired.

9.1 Confidence intervals for percentages and proportions

Where counts or scores have been accumulated as in a questionnaire, it is often desirable to obtain a confidence interval for a proportion or percentage of that score. Proportions follow the binomial rather than the normal distribution and this requires a modified calculation. The equation below can be used providing n is reasonably large (>50) and the proportion is not less than 0.1 or greater than 0.9 (10% and 90%). Outside this range, special tables must be consulted.

$$\text{confidence interval} = p \pm z \sqrt{(\hat{p}\hat{q}/n)}$$

where $\hat{p}$ is the proportion or percentage of interest and $\hat{q} = 1 - \hat{p}$.

BOX 9.3: CONFIDENCE INTERVAL CALCULATION FOR A PROPORTION

A sample of 100 houses was surveyed for wet rot (*Coniophora cerebella*) in floor timbers. It was found that 27% of the sample was infected. Find the 95% confidence intervals to this proportion.

Here, $\hat{p} = 0.27$ and $\hat{q} = 1 - 0.27 = 0.73$, giving

$$0.27 \pm \mathbf{1.96} \sqrt{(0.27 \times 0.73/100)}$$

$$= 0.27 \pm 0.087$$

$$= 27\% \pm 8.7\%$$

We can be 95% confident that the true (population) proportion lies within this interval. Again, by increasing the sample size, the confidence interval is reduced.

Key notes

- 95% confidence intervals to the population mean are obtained using the central limit theorem.

- Confidence intervals for a population mean are calculated by multiplying the standard deviation of the mean by an appropriate value of z (**1.96** for 95% limits) for large samples ($n \geq 30$) and an appropriate value of t for smaller samples.

- Another method is used to obtain confidence intervals of proportions.

Exercises

9.1 Obtain 95% confidence intervals for the following:

(i) $\bar{x} = 13.4$, $s = 1.56$, $n = 53$.
(ii) $\bar{x} = 12.1$, $s = 4.8$, $n = 18$.
(iii) $\bar{x} = 14.1$, $s = 1.8$, $n = 19$.
(iv) $\bar{x} = 5.1$, $s = 9.3$, $n = 12$.

9.2 Find (i) 95% and (ii) 99% confidence intervals to the population mean for the following measurements:

13.6   14.2   19.1   7.3   11.7   12.5   8.6   5.9   10.2   10.8

9.3 Find 95% intervals to the following proportions.

(i) $p = 0.38$, $n = 100$.
(ii) $p = 0.24$, $n = 100$.
(iii) 155 agreed, 345 disagreed.

CHAPTER TEN

Analysis of variance

In Chapter 8 several methods were given to compare two samples of measurements, with either a *t* test or a non-parametric test such as the Mann–Whitney test. Both of these tests are widely used, but they are limited in scope. Situations often arise where it is necessary to compare more than two samples. In such cases it is always possible to make *pairwise* comparisons. For example, if we took measurements of dissolved oxygen on three branches A, B and C of a river, we could compare A with B, A with C and then B with C. With a computer package the results could be obtained quickly, but as the sample number rises, the number of comparisons increases rapidly and even computer methods become tedious. There is also a fundamental objection to pairwise comparisons. Suppose there were seven samples which had all been randomly selected from a single normal population. A little arithmetic shows that a total of 21 pairwise comparisons are required. If you take the usual 5% probability level to reject the null hypothesis that the population means are equal, you are almost bound to find a significant difference between one of the pairs even though they were all drawn from the same population. Analysis of variance (ANOVA) overcomes this objection irrespective of sample size and employs a single test. It is also a versatile technique permitting the analysis of a wide range of sampling arrangements and may be used as an alternative to the *t* test. The simplest form of analysis of variance, known as one-way ANOVA, is explained below with an example. The null hypothesis is written as $\mu_a = \mu_b = \mu_c = \mu_d$ equality of the population means for four independent samples a, b, c and d.

The method will be illustrated with an example. In this case data from the pulping industry are analysed. The pulping of softwood for the making of paper is a huge international industry requiring large quantities of water. After felling, bark stripping and chipping, the wood is usually heated with

Table 10.1 Tonnes of water per tonne of paper pulped for four pulping mills in northern Europe

	a	b	c	d	
			Mill		
	22.0	18.6	23.1	20.3	
	18.8	16.0	18.0	18.5	
	19.5	17.7	20.0	19.6	
	23.7	14.6	21.0	22.8	
	21.9	19.9	23.7	23.8	
	19.9	18.6	22.4	21.4	Overall
n_i	6	6	6	6	24 (n)
$\bar{x}$	20.97	17.57	21.36	21.07	20.24
s^2	3.478	3.770	4.571	3.983	5.794
$\sum x_i$ (T_i)	125.8	105.4	128.2	126.4	485.8 ($\sum x$)
$(\sum x_i)^2/n_i$ ($= T_i^2/n_i$)	2637.6	1851.5	2739.2	2662.8	

strong reducing agents and alkali to release the wood fibres. The process is called pulping and is potentially highly polluting. The data in Table 10.1 are measurements taken randomly throughout the year of the amount of water used in the initial stage of pulping per tonne of wood at four pulp mills. These values are also plotted as histograms in Figure 10.1. The aim of the analysis is to determine if there is any significant difference between the four means. This is achieved using ANOVA.

Before commencing the analysis, look at the histograms in Figure 10.1. Two observations are worth noting. First, mills a, c and d are using similar amounts of water per tonne of pulp, while mill b seems to be using less. Second, all the sample variances appear to be about the same. The second observation can be confirmed from the calculated variances (s^2) shown below the data in Table 10.1. Here you will find other rows of figures used in the analysis. First, we need to take a brief sidestep to look at the variances a bit more. Notice that, for the combined data, the total variance is considerably higher than the individual sample variances. The latter are all between 3 and 5, while the former is almost 6. This increase is caused by differences between the individual sample means, and two sources of variation can be recognised, namely the random variation *within* each sample about the mean and the variation *between* the sample means. The *total* variation in the samples is the sum of the two

total variation = variation *between* samples + variation *within* samples

This relationship is used in ANOVA to partition the variance. If the samples are all drawn from the same normal population the two sources of variation (the *within* samples and *between* samples variances) will be the same. This is tested as the ratio of the two variances, which follows the 'F distribution'. This theoretical distribution, named after R. A. Fisher, differs

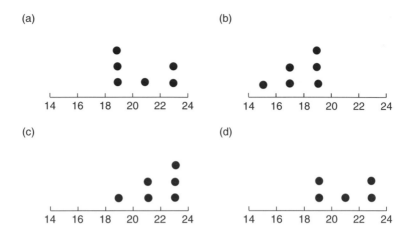

Figure 10.1 A dot plot for the ANOVA example.

from the t distribution as its shape is determined by *two* degrees of freedom. The distribution is positively skewed for most degrees of freedom and is therefore non-symmetric about the mode. In analysis of variance only the upper tail of the distribution is used in the critical tables.

BOX 10.1: ONE-WAY ANOVA CALCULATION

The first task is to obtain the *within* samples and *between* samples variances. The procedure is undertaken in a number of simple stages with the calculation of some sums of squared deviations. It is a good idea to set out a number of rows below the data, as shown in Table 10.1, which will be used in this example. First, place in one row the number of measurements in each sample, n_i, where the subscript i stands for the ith sample. For ANOVA, the values of n_i need not be all the same, though in this example they all happen to be 6. At the end of this row the total number of all measurements ($n = 24$) is shown. Next, place the $\sum x_i$ values in a row, i.e. the sum total for each column of measurements. The s^2 and $\bar{x}$ values provide useful information but they are not used in the initial calculations. Finally you need a row showing, for each sample, the quantity $(\sum x_i)^2/n_i$. For instance, the first sample has $\sum x = 125.8$. So the quantity required is $(125.8)^2/6 = 2637.6$. Having completed these rows, the sums of the squared deviations (usually abbreviated to 'sums of squares') can be calculated as follows.

The total sum of squares (SS_{total})

Combine all samples and calculate the sum of the squared deviations in the usual way

$$SS_{total} = \sum x^2 - (\sum x)^2/n$$

$$= 9970.18 - (485.8)^2/24$$

$$= 136.78$$

The between samples sum of squares (SS$_{between}$)

Add up the four $(\sum x_i)^2/n_i$ terms

$$2637.6 + 1851.5 + 2739.2 + 2662.8 = 9891.1$$

This sum is often written as $\sum T_i^2/n_i$, where $T_i = \sum x_i$. This is done to avoid using 'double sum' notation ($\sum\sum$), which many people find confusing.

Now subtract the 'correction term' $(\sum x)^2/n$

$$9891.1 - 485.8^2/24 = 9891.1 - 9833.4 = 57.77$$

This gives the 'between samples' sum of the squared deviations as 57.77.

The within samples (error, residual) sum of squares (SS$_{within}$)

Use is made of the relationship $SS_{total} = SS_{between} + SS_{within}$, which results in a simple subtraction

$$SS_{within} = SS_{total} - SS_{between}$$

$$= 136.78 - 57.77 = 79.01$$

Note that, as all sums of squares must be positive, the total sum of squares will have the highest value. In our example the between samples and within samples sums of squared deviations are similar in magnitude (57.77 and 79.01). However, for an ANOVA, we require sample variances and the sums of squares need to be divided by appropriate degrees of freedom.

Calculating the degrees of freedom and variance for ANOVA

Three degrees of freedom are required to accompany the sums of squares.

For SS_{total}, $df = n - 1$. For our example, $df = 24 - 1 = 23$.
For $SS_{between}$, $df = i - 1$. For the example, $df = 4 - 1 = 3$.
For SS_{within}, $df = $ total $df -$ between $df = 23 - 3 = 20$.

The variances are now calculated by dividing the sums of squares by the appropriate degrees of freedom. This step is completed in an ANOVA table.

One-way ANOVA table

Source of variation	SS	df	Mean square (variance)	F
Between samples	57.77	3	19.26	4.875*
Within samples	79.01	20	3.951	
Total	136.78	23	–	

The table provides a useful summary of the statistics. The variances, frequently referred to as the *mean squares*, are shown in the fourth column. The test statistic *F* is obtained as the quotient of the between and within samples variances (19.26/3.951).

5 PER CENT POINTS OF THE F DISTRIBUTION

$v_1 =$	1	2	3	4	5	6	7	8	10	12	24	∞
v_2												
18	4.41	3.55	3.16	2.93	2.77	2.66	2.58	2.51	2.41	2.34	2.15	1.92
19	4.38	3.52	3.13	2.90	2.74	2.63	2.54	2.48	2.38	2.31	2.11	1.88
20	4.35	3.49	3.10	2.87	2.71	2.60	2.51	2.45	2.35	2.28	2.08	1.84
21	4.32	3.47	3.07	2.84	2.68	2.57	2.49	2.42	2.32	2.25	2.05	1.81
22	4.30	3.44	3.05	2.82	2.66	2.55	2.46	2.40	2.30	2.23	2.03	1.78
23	4.28	3.42	3.03	2.80	2.64	2.53	2.44	2.37	2.27	2.20	2.00	1.76

Critical values of the F statistic are found in **Table VIII**, a small section of which is repeated above. You will find that each page of critical values refers to a single level of significance. Turn to the page showing the 5% level of significance. You will find the critical value of F by using the degrees of freedom for the SS_{within} and the $SS_{between}$; like a map reference. In the ANOVA table the respective df are 3 and 20 and the critical value is found in column 3 and row 20 as **3.10**. Turning to the 1% tables, the critical value will be found as **8.10**. The test value is thus exceeded only at the 5% level. The null hypothesis for the ANOVA states that the means of the populations are the same ($\mu_a = \mu_b = \mu_c = \mu_d$) and is rejected at the 5% level of significance. It is customary to employ a number of symbols in the ANOVA table to indicate the level of significance: * 5%; ** 1%; *** 0.1%; ns, not significant. If the F value is less than unity it is not usually entered as it cannot be significant. The tables are one-tailed because the quotient of the variances is specified.

Further testing for a significant ANOVA

If a significant difference is found, it is permissible to pursue the analysis further to determine which samples are significantly different from each other. An idea can be obtained by plotting the 95% confidence limits to the means as explained in Chapter 9. A better method is to perform the Tukey test explained in Box 10.2.

10.1 Assumptions for ANOVA

As a parametric test based on the normal distribution, certain assumptions apply to ANOVA. The measurements in each sample must be at least approximately normally distributed, and the samples should be homoscedastic (i.e. equal population variance for each sample) as in the case of the t tests. Normality can be tested using a stem and leaf plot or one of the other methods given in Chapter 7. Homoscedasticity can be checked by applying a simple test. Find the sample standard deviations for all i samples and then divide the smallest of these standard deviations by the largest. If the result is less than two the population variances may be considered homoscedastic. For the pulp mills example, the largest and smallest sample standard deviations were 2.137 and 1.865, respectively, giving a quotient of 1.146.

BOX 10.2: TUKEY TEST

Consult the half-**matrix** of statistics below. Each number in the table corresponds to the modulus (the positive value) of the difference in means of a sample pair. All combinations of samples are shown and are located by the table column and row. For example, the difference in the means of samples a and b appears top left as 3.40.

Half-matrix for the Tukey test on pulp mill data

	Sample a	Sample b	Sample c	Sample d
$(\bar{x})$	(20.97)	(17.57)	(21.36)	(21.07)
$\bar{x}_a - \bar{x}_b$				
Sample b	3.40			
Sample c	0.39	3.79		
Sample d	0.10	3.50	0.29	

The values in the above table are compared with a test statistic T, which must also be calculated from the data. It is obtained using the following relationship

$$T = q\sqrt{(\text{variance}_{\text{within}}/n_i)}$$

where q is obtained from **Table IX**, which requires information on the number, (a) of samples being compared (4) and the within samples degrees of freedom (20). This gives a value of 3.96 for q. The within samples variance is obtained from the ANOVA table as 3.951. The sample number (n_i) is 6. The test value for the example is

$$T = 3.96 \times \sqrt{(3.951/6)} = \textbf{3.213}$$

Finally, compare the values in the half-matrix above with the test value. If the test value is exceeded then the pair tested are significantly different in terms of their means at the 5% level. Thus the population means of samples a–b, b–c and b–d are significantly different, as is apparent from Figure 10.1. Note that the possibility remains that a wrong conclusion may be drawn when a large number of samples is compared and that the data sets must have equal numbers of measurements for the test.

One-way ANOVA with Minitab

One-way ANOVA with Minitab is illustrated with a further example. The data below are measurements of dibenzodioxin in the effluents from four pulp mills. These harmful chemicals form through the reaction of chlorine, used as a pulp bleach, with reactive aromatic compounds in the wood. The data from each mill are placed in columns 1–4 of a worksheet

C1	C2	C3	C4
1.41	0.48	1.02	1.06
2.05	0.86	1.32	0.95
0.78	1.34	0.68	1.73
1.82	0.59	0.54	0.51

Type C1–C4 in the 'Response in separate columns' box; then click 'OK'. The output is shown below.

```
Analysis of Variance
Source   DF    SS       MS       F      P
Factor    3   1.177    0.392    1.88   0.187
Error    12   2.503    0.209
Total    15   3.680
                                 Individual 95% CIs For Mean
                                 Based on Pooled StDev
Level    N    Mean     StDev    ----+---------+---------+---------+--
C1        4   1.5150   0.5569                    (---------*---------)
C2        4   0.8175   0.3832   (---------*---------)
C3        4   0.8900   0.3504    (---------*---------)
C4        4   1.0625   0.5045     (---------*---------)
                                 ----+---------+---------+---------+--
Pooled StDev = 0.4567            0.50      1.00      1.50      2.00
```

This table contains several new terms. *Factor* is used instead of *between samples* and *Error* has been substituted for *within samples*. The probability value (P) is given directly as 0.187. Since this value is greater than 0.05 the ANOVA is not significant.

Below the table, the population mean 95% confidence intervals are plotted for each sample. The pooled standard deviation listed in the printout is the $\sqrt{(\text{variance}_{within})}$, i.e. $\sqrt{(0.209)}$. The graph provides a good guide to significant differences beween samples in those cases when the ANOVA is significant. In this case the null hypothesis is accepted so there is no need to consult the plot.

10.2 The Kruskal–Wallis test

There is an equivalent non-parametric test for one-way ANOVA where the assumptions regarding the distribution of the data are more relaxed. The samples need only be independent of one another with the measurements at least at ordinal level. For this test the null hypothesis states that the sample populations are identical with respect to their medians. As with other non-parametric tests, ranking is involved and here all samples are first ranked together, as shown in Box 10.3, with the lowest rank assigned to the lowest value.

The example in Box 10.3 contains few data and it can be seen that the tables for the Kruskal–Wallis statistic (H) only extend to $i = 3$ and $n_i = 5$. For larger data sets an alternative table is needed. The distribution of H is close to that of χ^2 with df $i - 1$. For example, suppose that for $i = 5$ the value

BOX 10.3: KRUSKAL–WALLIS TEST

We shall look at the effects of bioremediation on an oil-contaminated beach using fertiliser application. The beach has been divided into 15 strips. Each strip has been randomly chosen and allocated 0 g m^{-2}, 5 g m^{-2} or 25 g m^{-2} ammonium phosphate solution to test whether naturally occurring oil-degrading bacteria are stimulated by the nutrient, leading to enhanced oil degradation. The amount of oil remaining after three months is expressed as a percentage of that originally present. In the table the columns in italics give the ranks of the measurements in the three treatments.

Bioremediation experiment using ammonium phosphate (NH$_4$)$_3$ PO$_4$

Control, 0 g m^{-2}	Rank	5 g m^{-2}	Rank	25 g m^{-2}	Rank
48	8	66	13	30	4
62	11	68	14	26	3
72	15	51	9	42	6.5
57	10	42	6.5	21	2
65	12	32	5	15	1
Rank sum R_i	56		47.5		16.5
R_i^2/n_i	627.2		451.25		54.45

The Kruskal–Wallis statistic (H) is calculated from the summed R_i^2/n_i term in the following way

$$H = \frac{12}{n(n+1)} \sum \frac{R_i^2}{n_i} - 3(n+1)$$

where the subscript i stands for the treatment (sample) number as before. For this example

$$\sum R_i^2/n_i = 627.2 + 451.25 + 54.45 = 1132.9$$

$$H = [12/15 \times (15+1)] \times 1132.9 - 3(16)$$

$$= 56.645 - 48 = 8.645$$

To test the significance of the statistic, turn to **Table X**, which provides exact probabilities for given values of H. The table only provides values for cases where $i = 3$ or less, which is the case here. The first three columns show the number of measurements in each sample. In our case the numbers were 5, 5, 5. Now consult the probabilities. When $H = $ **7.98**, $p = 0.01$, and when $H = $ **8.00**, $p = 0.009$. Clearly our test value of 8.645 has a probability of occurrence under H_0 of less than 0.009, so the null hypothesis must be rejected, i.e. the medians are significantly different.

of H had been calculated as 10.64. Go to the χ^2 critical tables (**Table VI**) with df $5 - 1 = 4$. You will find that the value of H has probability between 0.05 (**9.49**) and 0.025 (**11.14**), so in this case H_0 is rejected at $p = 0.05$ but not at $p = 0.01$.

One tied observation was encountered in the data. Unless the number of values tied exceeds 25% of all observations, the above method of calculating H is satisfactory. It is very unusual for the proportion of ties to exceed the above figure, but a correction is available which slightly increases the test value of H.

Kruskal–Wallis test with Minitab

The data must be arranged in two columns, C1 and C2. In the first column place the measurements and in the second the sample number. For example, for the following data, with three independent samples

Sample	1	2	3
	7.6	12.3	8.1
	8.4	7.4	0.9
	9.2	6.5	6.2
	8.1	8.2	3.7
		9.3	9.8

the Minitab worksheet is

C1	C2
7.6	1
8.4	1
9.2	1
8.1	1
12.3	2

and so on.

This kind of data arrangement is known as 'stacking' and can also be used for ANOVA.

On the menu bar:

Stat > nonparametric > Kruskal-Wallis

In the 'Response box' type C1, in the 'Factor box' type C2. Click 'OK'. The output appears below. In this case the null hypothesis is accepted.

```
Kruskal-Wallis Test on C1
     C2              N          Median       Ave Rank          Z
      1              4           8.250           8.6         0.64
      2              5           8.200           8.8         0.87
      3              5           6.200           5.3        -1.47
  Overall           14                           7.5
  H = 2.16        DF = 2      P = 0.340
  H = 2.16        DF = 2      P = 0.340      (adjusted for ties)
  * NOTE * One or more small samples
```

10.3 Two-way ANOVA

Two-way ANOVA is an extension of one-way anova permitting analysis along rows as well as columns. The model is quite different, however. In two-way ANOVA, the variance is partitioned into further components:

$$\text{total variation} = SS_{within} + SS_{betweenA} + SS_{betweenB} + SS_{interaction}$$

where A and B represent two variables. The interaction sum of squares is a special quantity, which is a measure of the dependence of one set of factors on another. For example, in a study involving two treatments, a and b, at two levels, 1 and 2, there are four combinations: a_1b_1, a_2b_1, b_2a_1 and b_2a_2. If the treatments are independent the effect of varying a_1 to a_2 will be the same in combination with either b_1 or b_2. It is not always possible to estimate the contribution of interaction to the total variation, but factorial experiments can be designed to account for it. Interaction becomes apparent in the phenomena known as **synergism** and **interference**. A synergistic effect is observed when the added separate effects of two variables are less than the combined effects. Synergism occurs in pollutant damage to plants where the effect of two pollutants combined is greater than the effect of the individual pollutants administered at the same concentration. Interference is the reverse, where the combined factors have less effect than the individual factors.

The statistical test for two-way ANOVA can be phrased thus: is there a significant difference between the means of variable A, and is there a significant difference between the means of variable B? It is best introduced as a simple example (see Box 10.4).

Two-way ANOVA using Minitab
The data must be arranged in three columns. In the first column place the measurements, in the second the column number and in the third the row number. For the electricity consumption example, the data would be input as follows:

C1	C2	C3
2.65	1	1
2.79	1	2
1.85	1	3
⋮		
3.09	2	1
2.95	2	2

and so on until all the data are entered.
    On the menu bar:

$$\text{Stat} > \text{ANOVA} > \text{Two way}$$

In the 'Response' box, put C1. In 'Row factor' type C2 and for the 'Column factor' type C3. Now click 'OK'. A two-way analysis of variance table will appear in the output.

BOX 10.4: TWO-WAY ANOVA

The measurements in the table are of electrical energy consumed in two sets of households over a year. A researcher wanted to answer two questions. First, is the amount of electricity consumed in the semi-detached properties the same as that consumed in terraced properties? Second, is electricity consumption the same in every month, or does it vary seasonally? The two sets of houses all had two working adult occupiers and had the same number of rooms of roughly equal size. A two-way ANOVA can help to answer these questions.

Total electricity consumption (MW month^{-1})

Month	Mid-terrace house	Semi-detached house	T_j	T_j^2/n_j
Jan	2.65	3.09	5.74	16.474
Feb	2.79	2.95	5.74	16.474
Mar	1.85	1.84	3.69	6.808
Apr	1.13	1.23	2.36	2.785
May	0.79	0.99	1.78	1.5842
Jun	0.82	0.81	1.63	1.3285
Jul	0.94	0.91	1.85	1.7113
Aug	0.77	0.70	1.47	1.0805
Sep	0.91	0.49	1.40	0.980
Oct	1.29	1.95	3.24	5.2488
Nov	1.88	2.01	3.89	7.5661
Dec	2.91	3.41	6.32	19.971
				$\sum T_j^2/n_j = 82.018$
T_i	18.73	20.38		
T_i^2/n_i	29.234	34.612		$\sum T_i^2/n_i = 63.846$

The data are laid out in a similar fashion to the one-way ANOVA example with some of the intermediate calculations listed in the right-hand column and in rows below the data. The total sum of the squared deviations is again calculated for the entire data set.

$$\text{total sum of squares } SS_{total} = \sum x^2 - (\sum x)^2/n$$

$$= 82.587 - (39.11)^2/24$$

$$= 82.587 - 63.733 = 18.8543$$

Remembering that $(\sum x)^2/n$ is the 'correction term' CT ($= 63.733$),

$$SS_i = \sum T_i^2/n_i - CT = 63.846 - 63.733 = 0.113$$

$$SS_j = \sum T_j^2/n_j - CT = 83.018 - 63.733 = 18.285$$

To complete the ANOVA, a value of SS_{within} must be obtained by subtraction using

$$SS_{total} = SS_{within} + SS_i + SS_j$$

Note that in this case the $SS_{interaction}$ cannot be calculated and is assumed to be zero. Thus

$$SS_{within} = SS_{total} - SS_i - SS_j$$

$$= 18.8543 - 0.113 - 18.285 = 0.4563$$

Degrees of freedom are calculated in a similar manner to one-way ANOVA

$$total\ df = n - 1 = 24 - 1 = 23$$

$$df_i = i - 1 = 2 - 1 = 1$$

$$df_j = j - 1 = 12 - 1 = 11$$

$$within\ df = total\ df - df_i - df_j = 23 - 1 - 11 = 11$$

The complete table appears below.

Two-way ANOVA table

Source	SS	df	MS	F
SS_i (houses)	0.113	1	0.113	2.72 ns
SS_j (months)	18.285	11	1.6622	40.1***
SS_{within}	0.4563	11	0.0415	
SS_{total}	18.854	23		

Consulting the critical values of F (**Table VIII**), the critical value of $F_{0.01,\ 1,\ 11}$ is **9.65**. The difference between the population means over the seasons is therefore highly significant, while the difference between the two types of houses is not. This becomes evident if the data are plotted as histograms.

The same assumptions concerning normality and homoscedasticity apply to this ANOVA but, with only two columns of data, testing for normality is not possible for factor j.

There is also a non-parametric test equivalent to this ANOVA, the Friedman test. It is a ranking procedure and is available on Minitab.

Key notes

- One-way analysis of variance (ANOVA) can be used to compare the means of $i = 2$ to many samples.

- All methods of ANOVA partition the sample variance into two or more additive quantities.

- Pairwise comparisons between the samples can be made using the Tukey test, providing the ANOVA indicated a significant difference between the means.

- ANOVA should be performed on normally distributed, homoscedastic samples.

- The Kruskal–Wallis test provides an alternative to one-way ANOVA with relaxed assumptions of normality.

- Two-way ANOVA can be used to analyse samples which have been subjected to two levels of treatment.

Exercises

10.1 Six measurements of chlorine (ppm) were taken randomly at six sampling stations in a water distribution system. Use one-way ANOVA to determine if the chlorine concentration changes significantly along the system.

Station	1	2	3	4	5	6
	1.52	1.54	1.48	1.46	1.32	1.27
	1.43	1.46	1.42	1.51	1.29	1.36
	1.46	1.50	1.51	1.49	1.48	1.46
	1.56	1.49	1.53	1.51	1.50	1.48
	1.50	1.42	1.39	1.43	1.30	1.41
	1.48	1.41	1.45	1.43	1.38	1.49

10.2 Road surfaces are known to affect fuel consumption and an experiment was performed using five identical vehicles on six road surfaces, A–F. The distance travelled by each vehicle was then recorded as kilometres per litre of fuel. Use one-way ANOVA to determine if the mean petrol consumption differed significantly between the road surfaces.

A	B	C	D	E	F
12.21	12.05	13.35	10.46	12.53	12.95
12.80	11.06	10.98	10.32	12.77	13.01
11.96	10.98	12.86	9.89	12.08	12.88
11.70	11.85	12.99	11.83	12.05	12.65
12.04	11.63	12.78	12.01	12.67	12.03

10.3 Seven samples of soil were collected randomly from each of five fields, A–E, as part of a metal contamination study. Each soil was dried, digested and analysed for lead (Pb). The results are shown below. Perform a one-way ANOVA to determine if the concentration of lead differs in the five fields.

Pb concentration in soils (ppm)

A	B	C	D	E
6.3	5.0	4.2	6.1	8.3
6.8	5.9	4.1	7.2	8.2
6.5	5.8	4.3	7.5	8.2
6.0	6.1	4.0	7.5	8.2

5.3	6.8	4.5	7.0	8.1.
5.9	6.7	3.9	6.1	8.2
6.1	6.7	4.5	6.9	8.1

10.4 A bacteriological survey of seawater taken from five sandy bays, A–E, was undertaken. Five random samples were taken from each bay and viable counts of *Escherichia coli* were determined. Use the Kruskal–Wallis test to determine if there is a significant difference in bacteria numbers. Note that bacterial counts are rarely normally distributed so the use of ANOVA is questionable in this case.

Numbers of *E. coli* in 100 ml seawater

A	B	C	D	E
15	1300	12 200	750	10
560	9800	17 400	17 200	100
7900	1800	1200	1000	2300
500	350	4900	50	2800
13 200	850	7900	5600	700
	60			

10.5 In a study of acid rain effects on UK soils, the pH of four soils was determined on six random samples from each of four sites, A–D. Use the Kruskal–Wallis test to determine if there is a significant difference in the soil pH at the four sites. Note that pH is a logarithmic quantity which would not be expected to follow a normal distribution.

A	B	C	D
6.42	7.01	6.95	7.45
5.95	6.98	7.14	7.02
5.80	6.86	7.27	6.99
6.06	7.23	6.23	6.85
6.15	6.41	6.11	6.03
5.92	6.81	6.00	7.02

10.6 In a study of nitrogen dioxide pollution by vehicle emissions, 12 models of car were driven for 1 h in a town area (A) and an adjacent country area (B). Each car interior carried a triethanolamine disc, which adsorbs nitrogen dioxide, permitting the mean nitrogen dioxide concentration to be calculated over the 1 h period. Perform a two-way ANOVA to determine whether the pollutant concentration differs significantly between vehicle model and location of the drive.

Vehicle model	NO_2 concentration (ppb)	
	Area A	Area B
1	17.4	12.3
2	21.8	18.2
3	42.3	21.8
4	38.1	41.0
5	21.6	16.4

6	36.0	19.8
7	39.9	37.3
8	42.2	22.1
9	18.6	18.8
10	29.8	9.4
11	15.7	10.6
12	14.2	11.2

10.7 As part of an air pollution study, visibility in air was measured over a town over a one-year period. Measurements were made at two-monthly intervals at six heights above the ground as shown below. Perform a two-way ANOVA to determine if visibility varied (a) with time and (b) with height above the ground.

Visibility in town air (km)

Height above ground (m)	Months 1–2	Months 3–4	Months 5–6	Months 7–8	Months 9–10	Months 11–12
0	13.5	12.8	14.2	2.8	4.3	13.5
10	17.2	13.0	14.0	3.1	4.5	16.8
20	18.1	16.2	14.7	6.2	5.6	21.6
30	14.3	15.4	12.6	4.5	9.7	9.2
40	10.6	19.0	22.8	8.6	6.5	8.8
50	19.5	19.3	7.6	10.1	8.0	7.0

CHAPTER ELEVEN

Testing if a relationship occurs between two variables using correlation

In research, we are often interested in the way one variable responds to changes in another variable. For example, one of our undergraduates was interested in growth of the lichen *Evernia prunastri* in relation to air pollution. This lichen grows on the bark of trees and she measured the maximum length of the lichen and the distance of the lichen from a town centre. Twenty sites, which ranged from 0 to 10 km from the town centre, were chosen and the results were plotted as a scattergraph (Figure 11.1).

It is possible to deduce some interesting information from the graph. To begin with, a closed line surrounding all the points would be approximately elliptical in shape. This suggests that, although there is considerable variation in the data, there is a trend of decreasing thallus size as the town centre is approached. If we assume that thallus size is an indicator of the health of this lichen, then it appears that the growth of *Evernia* becomes stunted as the town is approached. We can also identify one point on the graph that lies at a considerable distance from the others. Such points are termed **outliers** or 'wildshots'. There is also a gap in the data between 4 and 5.5 km where there were no trees supporting the growth of the lichen.

It is known from previous work that *Evernia* is sensitive to atmospheric sulphur dioxide pollution and the student's results are not surprising. However, if we knew nothing of this lichen, the results would benefit from further analysis. A statistical technique termed **correlation** enables us to quantify the relationship between two variables such as *Evernia* length and distance from the town centre. It involves the calculation of a coefficient which ranges from +1 to −1. Unlike most other statistical quantities, knowledge of the coefficient alone tells us something about the relationship between the two variables. A formal statistical procedure is also available, which provides us with a rigorous test of the association between the variables.

Reference to Figure 11.2 enables us to see more clearly how the value of the correlation coefficient relates to particular scattergraphs. In Figure

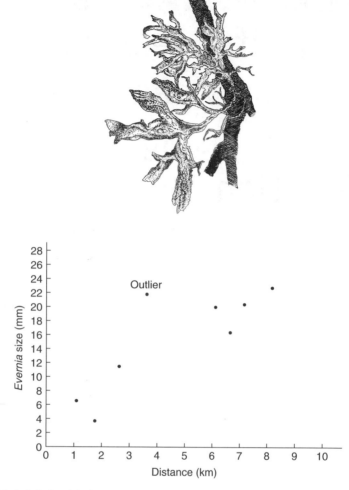

Figure 11.1 Relationship between *Evernia* length and distance from a town centre.

11.2(a) it is apparent that a near 'perfect' relationship occurs between the variables with a correlation coefficient approaching +1. As the scatter in the data increases the coefficient begins to fall (Figure 11.2(b)). In these diagrams, as one variable increases, so does the other, and in such cases the correlation is said to be positive. However, we often find cases where one variable decreases as the other increases, as shown in Figure 11.2(c), and the correlation is then described as negative. When the value of the coefficient, which is given the symbol r by statisticians, falls below 0.5, trends become obscure even when we have a large number of measurements (Figure 11.2(d)).

Before pursuing correlation any further we need to look at an important term, the *covariance,* a quantity from which the correlation coefficient is obtained. The covariance is a measure of variance and is a sum of squared

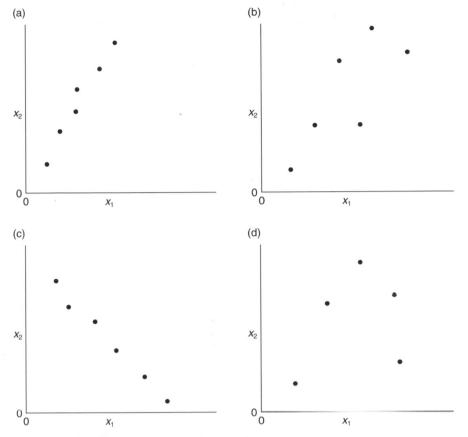

Figure 11.2 Scattergraphs with contrasting correlation coefficients (see text).

deviations divided by degrees of freedom. It involves a sum of products – a computation not previously encountered. Look at Figure 11.3, which provides a scatterplot for variables x_1 and x_2. These may have been the *Evernia* length and distance or any other variable pair. Which variable is called x_1 and which x_2 is immaterial providing we label them for purposes of identification. In Figure 11.3 the three points are labelled A, B and C, corresponding to three measurements of both x_1 and x_2. Also shown in the figure is a point representing the mean of the two variables, marked as a bold x and termed the **centroid**. The covariance is obtained by first measuring the abscissa and ordinate of each point from x, shown by the broken lines. These distances can be seen to represent the difference between an x_1 and x_2 value and the mean. If the difference is negative, the negative sign is retained. The two perpendicular distances are then multiplied together and summed. The result could be either positive or negative depending on the positions of the points in relation to the centroid.

For example, in Figure 11.3 the summed products are

$$(a \times -b) + (c \times d) + (-f \times -e)$$

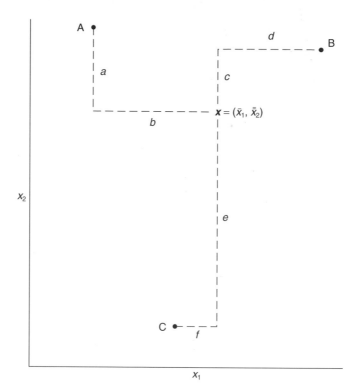

Figure 11.3 Diagram illustrating the meaning of covariance.

If $a = 4$, $b = 6$, $c = 3$, $d = 5$, $e = 10$ and $f = 2$ the sum of the products is $-24 + 15 + 20 = 11$. To obtain the covariance we now divide this sum by $n - 1$ degrees of freedom, giving a value of 5.5. This covariance is positive, indicating a positive association between x_1 and x_2 and showing that there is a tendency for x_1 to increase as x_2 increases. If the covariance had been negative then, as x_1 increased, x_2 would tend to decrease. However, with only three datum points this cannot be appreciated in the figure. The covariance can be written formally as

$$\text{covar}(x_1,\ x_2) = \frac{\Sigma(x_1 - \bar{x}_1)(x_2 - \bar{x}_2)}{n - 1}$$

11.1 Product–moment correlation

The *product–moment* correlation coefficient is obtained by dividing the covariance by a pooled standard deviation ($s_{x_1} s_{x_2}$). This is done to make the correlation independent of the original scale of the measurements being taken, making it more practicable. Unlike most statistical quantities, the correlation coefficient (r) takes values in the range -1 to $+1$, from perfect negative to perfect positive correlation. The coefficient is sometimes called Pearson's r.

BOX 11.1: PRODUCT–MOMENT CORRELATION CALCULATION

Calculating the product–moment correlation coefficient (r) by hand is straightforward when the number of measurements is below 20, but soon becomes tedious as the number of measurements increases. The formula to calculate r is shown below. This is one of the convenient forms of the equation and embodies the covariance divided by the pooled standard deviation. The expression in the numerator is called the 'sum of products'.

$$r = \frac{\sum x_1 x_2 - (\sum x_1 \sum x_2)/n}{\sqrt{\{[\sum x_1^2 - (\sum x_1)^2/_n][\sum x_2^2 - (\sum x_2)^2/n]\}}}$$

For the calculation we shall look at some of the *Evernia* data shown in the table.

Preliminary calculations for the product–moment correlation coefficient

x_1, distance from town (km)	x_2, Evernia length (mm)	x_1^2	x_2^2	$x_1 x_2$
3.16	11.6	9.986	134.56	36.656
6.10	20.1	37.21	404.01	122.61
7.15	20.8	51.126	432.64	148.72
1.76	3.8	3.098	14.44	6.688
3.61	18.0	13.032	324.0	64.98
8.18	23.2	66.912	538.2	189.8
9.31	27.3	86.68	745.29	254.16
6.64	16.15	44.090	260.82	107.24
1.05	3.80	1.103	14.44	3.99
6.90	11.1	47.61	123.21	76.59
$\sum$ 53.86	155.85	360.85	2991.6	1011.43

$n = 10$

The scattergraph in Figure 11.1 suggests a positive correlation. The quickest way to calculate the $\sum x_1 x_2$ term, assuming that your calculator has no xy function, is to multiply the first two terms across (i.e. 3.16×11.6) and store the result (36.656) in the memory. Now multiply the second pair ($6.10 \times 20.1 = 122.61$) and add it to the product in the store and store the new product sum ($122.61 + 36.656 = 159.266$). Continue until all the products are summed.

$$r = \frac{1011.43 - (53.86 \times 155.85)/10}{\sqrt{\{[360.85 - (53.86)^2/10][2991.6 - (155.85)^2/10]\}}}$$

$$= \frac{172.02}{\sqrt{(70.76 \times 562.68)}}$$

$$= 172.02/199.54$$

$$= 0.862$$

The resulting value of r demonstrates a positive correlation between the variables. The **numerator**, which contains the covariance term, must be positive when x_1 increases with x_2. The denominator should never be negative as it is a standard deviation.

11.1.1 Testing the significance of r

It is usual to determine if the correlation coefficient (r) could have come from a population whose parametric correlation coefficient (ρ) is zero. If this is the case then the two variables are uncorrelated. The null hypothesis takes the form $H_0 : \rho = 0$.

For most purposes, hypothesis testing is achieved by consulting critical values of r (**Table XI**) with $n - 2$ degrees of freedom. If the modulus of r is greater than the tabulated value for the specified p value, the null hypothesis is rejected.

In the above example, the test value of r was found to be 0.862. Since $n = 10$, the degrees of freedom, $n - 2 = 8$. Consulting **Table XI**, the critical value of r for the 5% level of significance is **0.632**. The value at the 1% level is **0.765**, so we reject H_0 at both the 5% and the 1% level of significance. There is therefore strong evidence for a positive association between *Evernia* size and distance from the town centre.

Look again at **Table XI**. You will see that when the degrees of freedom exceed 30, only selected values of r are provided. For example, if $n = 65$, then $df = 63$, but there are no critical values available in the table for 63 degrees of freedom. In such cases, it is simplest to interpolate the critical values on either side of $df = 63$ (see Chapter 8).

If critical values of r are not available you can use the relationship below, which transforms your sample r to Student's t

$$t = r \sqrt{[(n - 2)/(1 - r^2)]}$$

Suppose for $n = 65$ that your sample r was -0.216. Using this equation

$$t = -0.216 \times \sqrt{[63/(1 - 0.046\,66)]}$$

$$= -0.216 \times \sqrt{(66.083)}$$

$$= -1.756$$

Since your critical t values also 'jump' degrees of freedom above 30, you still need to interpolate for an accurate value, but you do have more significance levels with which to test H_0. For $df = 63$ the 5% critical value of t (two-tailed) using linear interpolation is **1.999** (**Table III**) so H_0 is accepted. In a case such as this, where the sample value clearly lies in the acceptance region, an interpolation was not really needed to complete the test.

Significance testing of r usually takes the two-tailed form and the critical tables for r are two-tailed. One-tailed procedures are occasionally used where the alternative hypothesis may be $H_1 : \rho > 0$ rather than $H_1 : \rho \neq 0$. In such cases, the 10% significance values are consulted for $p = 0.05$ as before. Here, too, the t transformation will prove useful.

Correlation coefficients are used to measure the strength of association between a pair of variables and to test whether the association is greater than can be expected by chance. However, it is dangerous to read too much

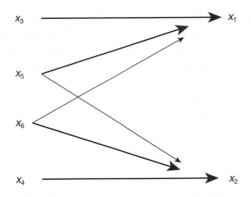

Figure 11.4 Illustration of causation in correlation.

into a significant correlation. For instance in Europe following the Second World War a significant correlation was found between the per capita birth rate and the number of stork's nests. The reason for the birth rate increase was complex and related to socioeconomic factors and there is no obvious relationship to the start population. A large number of factors was responsible for the birth rate increase and to choose just one would be foolhardy without a detailed analysis of the phenomenon. Situations where two variables are totally correlated are rare with continuous variables, but they sometimes arise if discontinuous variables are involved. An example is provided by the dust layers found in glacier ice. Each year, a layer of dust settles on the glacier during the summer when snowfall is reduced. Thus a perfect positive correlation ($r = 1$) exists between the number of dust layers and time measured in years.

In most situations we have a case rather like that idealised in Figure 11.4, where two variables, x_1 and x_2, have several causative factors influencing them. Variable x_3 directly affects x_1, but does not affect x_2. Variable x_4 likewise influences x_2 but not x_1, while variables x_5 and x_6 influence both x_1 and x_2 but to different degrees. To analyse the relationships in depth would require many experiments. They cannot be determined from the correlation coefficients alone.

Unlike regression (Chapter 12), confidence intervals for correlations cannot be obtained easily as we only have a scatter of points rather than a line to work with. However, it is possible to obtain 'confidence regions' and the reader is referred to Sokal and Rohlf (1995) for more information. It is quite common to see r values quoted as part of a regression analysis, but here it should not be considered as a sampling statistic as the data for model 1 regression will not follow the bivariate normal distribution.

11.2 Assumptions

As a parametric test based on a bivariate normal distribution, certain assumptions apply regarding product–moment correlation. Both sample distributions should be normal and the measurements should be taken at interval level.

If these criteria are not fulfilled a transformation (Chapter 7) might be applied, but it would be better to use a ranked correlation coefficient.

Always draw the scattergraph before analysing the data. Any apparent curvature in the distribution of points will suggest a significant departure from linearity and product–moment correlation is inappropriate. Sample size is also worth considering. Unless strong correlations occur, a sample size less than ten is insufficient to detect departures from normality. However, analysing a larger sample ($n = 50$) compared with a smaller sample ($n = 20$) will make little difference to the value of r, but would be worthwhile if the initial correlation with $n = 20$ proved to be only marginally significant.

The correlation coefficient is used extensively in environmental research, where the observer has no control over the variables as in field work, and the social sciences. *Autocorrelation* is often employed in time series analysis.

Product–moment correlation with Minitab
The two sets of data are input into the first two columns, C1 and C2, of the worksheet as shown in the example below.

```
C1   1.7  7.4  8.3  6.7  7.6  8.8  9.8  8.8  11.2  5.5  6.3
C2   2.3  8.1  8.4  9.1  9.0  9.7  9.9  8.6  11.6  7.4  8.4
```

Before the analysis, have a look at the scattergraph with *Graph > plot*. Input C1 in the 'X' box for 'Graph 1' and C2 in the 'Y' box and then click 'OK'. Since the data show an approximately ellipsoidal scatter they may be assumed to be normally distributed and the correlation coefficient can be obtained:

On the menu bar:

Stat > Basic statistics > correlation

In the dialog box, place C1 C2 in 'Variables' and click 'OK'. The output showing the probability under H_0 is given below. In this case the correlation coefficient is positive and highly significant.

```
Worksheet size: 100000 cells
Correlation of C1 and C2 = 0.936, P-Value = 0.000
```

11.3 Spearman's rank correlation

Spearman's test provides a non-parametric correlation coefficient which also ranges from +1 to −1. It can be used where the product–moment test is unsuitable or as an alternative procedure. Along with other non-parametric tests, ranking is involved and the method is modified where there are tied observations. The null hypothesis states that the two variables are not associated in the population, so that the test value of r differs from zero only by chance.

BOX 11.2: SPEARMAN'S TEST

In the UK a qualitative scale is available for the assessment of sulphur dioxide pollution using the presence of certain corticolous (tree-inhabiting) lichens. The scale ranges from 0 to 10. Scale 0 is used for trees completely free of lichens, indicating a high level of pollution; while scale 5 is applied to trees with a range of moderately SO_2-intolerant species, including *Evernia prunastri* and *Hypogymnia physodes*. The scale is believed to have an approximately linear response to SO_2 concentration in the air.

The measurements in the table provide mean winter levels of smoke (μg m^{-3}) for a range of UK towns. Since coal fires produce both smoke and sulphur dioxide, it would be interesting to see if there was a significant association between the lichen scale and smoke concentration. Because the lichen scale is ordinal, we use a Spearman's test.

Relationship between average smoke concentration and corticolous lichen zone

1	2	3	4	5	6
Smoke (μg m^{-3})	Lichen zone	Smoke rank	Lichen rank	Rank difference d	d^2
89	0	8	1	7	49
29	9	3	8	−5	25
43	3	6	3	3	9
102	4	9	4	5	25
32	6	4	5	−1	1
85	1	7	2	5	25
22	7	2	6	−4	16
33	8	5	7	−2	4
20	10	1	9	−8	64

$n = 9$

The first thing to do is to draw a scattergraph, which suggests a strong negative correlation between the variables. The null hypothesis in this case states that there is no association between the two variables.

Set out the data as shown in the table. Now rank the two variables giving the smallest value the lowest rank. It is important to rank the variables separately so as to produce two sets of ranks, one for each variable. These appear in columns 3 and 4 of the table.

Obtain the difference between the two sets of ranks and place it in column 5. Finally, each difference is squared and the result is placed in column 6. Since the differences are squared, it does not matter which way the difference is taken. For example, in row 1 the difference (d) is either $8 - 1$ or $1 - 8$. The squared result (d^2) is 49 either way. The next job is to sum the d^2 values to obtain Σd^2 as 218.

The formula for Spearman's rank correlation coefficient is

$$r = 1 - \frac{6\Sigma d^2}{n^3 - n}$$

This gives

$$r = 1 - 6 \times 218/(9^3 - 9)$$

$$= 1 - 1308/720$$

$$= 1 - 1.8167$$

$$= -0.917$$

For the test value we take the modulus of r, i.e. $|r| = 0.817$. Now turn to the critical values of Spearman's r (**Table XII**). For $n = 9$, $r = \mathbf{0.733}$ at $p = 0.01$. Therefore the test value exceeds the two-tailed critical value and we reject the null hypothesis. There is evidence for a negative association between the lichen scale and smoke concentration.

11.3.1 Tied values

We were lucky in the example of Box 11.2 not to encounter ties, since they are likely to occur when an integer variable ranges from 0 to 10. If a small number of ties had occurred then we would treat the ranks in the normal way by averaging the ranks for the tied values and then continue the calculations for Σd^2 and r as before. However, if the number of tied values is large with respect to n, we need to use a modification of the equation. You should use the method below if the number of ties (in *either* sample) is equal to or greater than 50% of the total number of measurements. This is because the quantity $(n^3 - n)/12$ is used as an estimate of a sum of squared deviations. If many ties are present in one (or both) ranked variables, the relationship no longer holds and we need a new expression. This is given below.

Spearman's rank correlation (r) with ties is given by

$$r = \frac{\Sigma x^2 + \Sigma y^2 - \Sigma d^2}{2\sqrt{(\Sigma x^2 \Sigma y^2)}}$$

where

$$\Sigma x^2 = (n^3 - n)/12 - \Sigma T_x$$

and

$$\Sigma y^2 = (n^3 - n)/12 - \Sigma T_y$$

$T_x = (t^3 - t)/12$, where t is the number of tied variates in a tied group for variable x. Each tied pair is assigned a value of $t = 2$ and t is then used to calculate values of T_x or T_y. The subscripts x and y are used to refer to the two variables used in the calculation. If only one tied pair had been encountered in the ranked scores of variable x, then $t = 2$ and $T_x = (2^3 - 2)/12 = 0.5$. If three values formed a triplet (i.e. all three with the same rank), then $t = 3$ and $T_x = (t^3 - t)/12 = (27 - 3)/12 = 2$. The T_x values are then

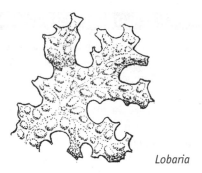

Lobaria

summed in cases where there is more than one tied group per variable to give ΣT_x.

The use of x and y in this test rather than x_1 and x_2 might appear confusing. It is used to make the formula for calculating r easier to read.

Spearman's test with Minitab
Enter the data into the first two columns, C1 and C2, of a worksheet. Before the coefficient is calculated the data must be ranked and the ranks must be placed in two more columns, e.g. C3 and C4.

On the menu bar:

Manip > Rank

In the 'rank data in' box enter C1 and in the 'store ranks' box enter C3 and click OK. Perform the same operation with C2, storing the ranks in C4.

Now return to the menu bar:

Stats > Basic statistics > Correlation

In the 'Variables' box type C3 C4. Click 'OK' and the rank correlation coefficient with its p value is obtained.

Key notes

- Correlation is a technique used to clarify the relationships between two variables.

- For a product–moment correlation the variables are assumed to follow the bivariate normal distribution.

- The correlation coefficient is a numerical measure of the strength of the relationship between two variables, x_1 and x_2.

- All correlation coefficients range from +1 to −1.

- The Spearman rank correlation coefficient is used for non-normally distributed ordinal or interval/ratio measurements.

- Always draw the scattergraph before obtaining a correlation coefficient.

BOX 11.3: SPEARMAN'S TEST WITH A LARGE NUMBER OF TIES

This example introduces a study involving the lichen *Lobaria pulmonaria,* a large leafy species which is very sensitive to pollutants and forest disturbance. It is now rare or extinct from large areas of Europe.

A study was performed to see if the relative abundance of *Lobaria* could be related to forest age. For this a scale was drawn up to account for the frequency and state of health of *Lobaria* scored on a scale 1–10, the lowest value indicating the least healthy. The age of the forest was estimated from the presence of a number of other old forest 'indicator plants'. Plotting the data failed to show any obvious correlation between forest age and *Lobaria* health.

Data for *Lobaria* health vs forest age

x, forest estimated age (yr)	y, Lobaria status	Forest rank	Lobaria rank	d	d^2
70	3	2	5.5	−3.5	12.25
100	1	4	1.5	2.5	6.25
300	1	10	1.5	8.5	72.25
270	2	9	3.5	5.5	30.25
80	5	3	7	−4	16.0
150	8	5	10	−5	25.0
50	2	1	3.5	−2.5	6.25
320	3	11	5.5	5.5	30.25
170	6	7	8	−1	1.0
190	9	8	11	−3	9.0
160	7	6	9	−3	9.0

$\sum d^2 = 217.5$ $n = 11$

There are three sets of ties in the *Lobaria* data and in each case two values are tied together (i.e. $t = 2$), giving $T = (2^3 - 2)/12 = 0.5$ for all three groups. Thus $\sum T = 0.5 + 0.5 + 0.5 = 1.5$.

$$\sum x^2 = (n^3 - n)/12 - \sum T_x = (11^3 - 11)/12 - 0 = 110$$

$$\sum y^2 = (n^3 - n)/12 - \sum T_y = (11^3 - 11)/12 - 1.5 = 108.5$$

The top line (numerator) of the equation for r is

$$110 + 108.5 - 217.5 = 1$$

The denominator is given by

$$2\sqrt{(\sum x^2 \sum y^2)} = 2\sqrt{(108.5 \times 110)} = 218.49$$

Hence

$$r = 1/218.49 = 0.005$$

Consulting the critical values ($p = 0.05$) for Spearman's r (**Table XII**) for $n = 11$ gives r (two-tailed) = **0.618**. Our test value is close to zero, so there is no support for a significant correlation between the two variables. Note that in this case, if there was strong evidence from earlier work for a positive association between *Lobaria* status and forest age, a one-tailed test would have been more appropriate. However, in this example the conclusions would have been the same.

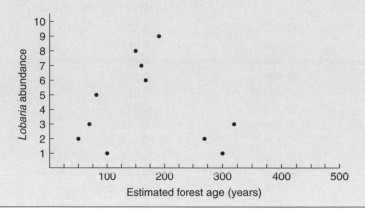

Exercises

11.1 The data below are for Secchi disc transparency and phytoplankton production in 15 Welsh lakes. Obtain the product–moment correlation coefficient and determine its significance.

Secchi disc transparency (m)	Phytoplankton production (mg C m^{-3} h^{-1})
7.95	390
7.25	235
6.65	590
8.05	605
4.35	710
3.55	335
5.95	360
6.45	205
5.75	770
4.45	650
0.65	820
9.05	110
6.65	450
3.50	520
1.85	560

11.2 Measurements were made on 17 sewage-contaminated seawater samples for total viable bacteria and ultraviolet absorbance (a measure of dissolved organic matter). Obtain the product–moment correlation coefficient and its significance after transforming the bacterium counts to a log-scale (use $\log_{10}$ counts).

Bacteria per 100 ml water	UV absorbance (350 nm, 100 mm path)
331 000	0.020
282 000	0.014
35 500	0.022
54 900	0.025
5485	0.029
36 700	0.030
30 200	0.043
5370	0.043
177 800	0.051
33 100	0.050
9120	0.053
1090	0.050
41 700	0.071
3020	0.062
1660	0.072
570	0.072
280	0.089

11.3 Determine whether there is a significant correlation between phosphoric acid production and phosphate disposal to sea in nine European countries.

Phosphoric acid annual production (kt)	Phosphate annual disposal to sea (t)
243	20
65	490
199	520
240	670
430	3440
470	5
180	45
45	15
491	3390

11.4 A study was undertaken on 17 30-year-old men to determine if there was a relationship between levels of arsenic in their fingernails and in their hairclippings. Obtain Pearson's correlation coefficient and determine whether the variables are significantly correlated.

Arsenic concentration (ppb)	
Fingernail	Hair
5.3	8.1
9.8	3.9
14.7	34.1
17.6	25.8
10.6	5.7
10.2	30.0
2.7	11.3
5.1	27.3
11.6	18.0
12.6	27.7
17.4	32.2
2.8	5.8
5.4	17.3
14.2	24.3
7.7	14.9
1.7	3.8
17.1	37.6

11.5 A study was made of paint discoloration in relation to the amount of exposure to sunlight. Both variables were recorded on a relative scale and the results are tabulated below. Perform a Spearman test to find out whether discoloration is significantly correlated with light exposure.

Paint discoloration (discoloration increases with the scale)	Sunlight exposure
45	32
110	61
103	46
41	71
156	92
95	27
96	55
60	20
165	75
155	53
115	72
10	11
141	84
107	51
145	62
195	77
167	26
69	33
72	38
35	30

Mathematical description of the relationship between two variables using regression

We saw in the preceding chapter how it is possible to obtain a measure of correlation between two variables. In this chapter another approach is pursued in order to provide an estimate of one variable from another using a simple mathematical expression. Such estimates are of great value in environmental modelling where it is possible to predict the likely outcome of events given sufficient quantitative knowledge of the processes involved.

Two regression models, model 1 and model 2, can be used depending on the nature of the data. Situations often arise in experiments where we deliberately control the conditions to reduce the variation to a small number of factors. For example, an experimenter may wish to study the rate of disappearance of a pesticide in a seawater sample. A known concentration of the pesticide would be added to the seawater and then after fixed intervals of time a small subsample would be taken and the concentration of the pesticide determined chromatographically. In such an experiment, the irradiance, temperature, salinity and pH would be held constant, because the decay of the pesticide may be altered by these factors.

Such types of experiment are commonly undertaken so that a response can be measured as a function of a single variable (in this case time). Model 1 regression analysis is well suited to these types of experiments because, as in most statistical techniques, there are conditions attached to the procedures. One of these conditions states that the x variable, often referred to as the **independent** variable, must be measured 'without error'. This sounds like an impossible condition, but in practice it means that the variable must be measured with a high degree of accuracy and is not subject to random variation.

The **dependent** variable (y) may vary randomly and its 'error' should follow a normal distribution. Other assumptions concerning linearity and homoscedasticity will be considered later. As a general rule, when experiments involve one variable, which is under the control of the investigator (e.g. time, temperature), a model 1 regression will be appropriate. Where neither variable is under control and both are subject to random variation, a model 2 regression should be used. Model 2 regression will often be needed in field studies where conditions are difficult to control.

Before we look at regression analysis in detail, it is important to understand how straight lines are drawn on graph paper using a simple regression equation.

12.1 The straight line equation

Ordinary graph paper provides us with two axes, usually referred to as the abscissa or x axis, which runs across the page, and the ordinate or y axis, running down the page. The two axes cross at the **origin** where the x and y values are both zero.

Any straight line can be drawn on ordinary graph paper with two pieces of information. First we need to know where the line intersects the ordinate (Figure 12.1(a)) and second we must know the slope (gradient) of the line. The point where the line crosses the ordinate is called the **intercept**. If the line intersects above the x axis (for $y = 0$) then the intercept is reckoned positive, and if below, it is negative. The gradient of the line is calculated as the ratio c/d in Figure 12.1(a). Steep gradients will have a ratio that exceeds unity, while gentle gradients are less than unity. A positive gradient is obtained when the x values increase with the y values and a negative gradient when the y values decrease when x values increase. A line running parallel to the x axis has a gradient of zero. A vertical line would have a gradient of infinity and a line that passes through the origin has an intercept of zero.

Any straight line can be described by an equation of the form

$$y = a + bx$$

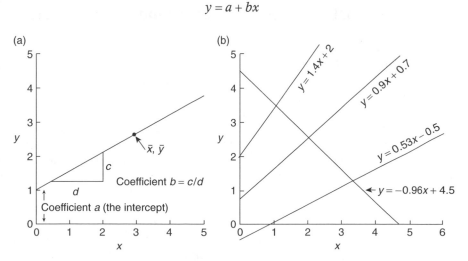

Figure 12.1 (a) A regression line with regression coefficients a and b; (b) some examples of regression lines with their equations.

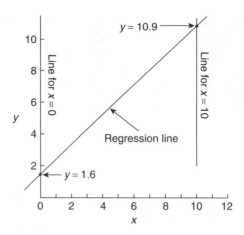

Figure 12.2 Method for plotting the regression line.

where a is the intercept and b the gradient. Several straight lines and their equations are shown in Figure 12.1(b).

A straight line defined by $y = 1.6 + 0.93x$, for example, can be drawn on graph paper by finding two values of y from two values of x. A convenient value of y is found by letting $x = 0$, so that y becomes equal to the intercept, which is 1.6 in this case. Now let $x = 10$ (or any other value that fits onto the graph paper, but not too close to the other point); then $y = 1.6 + 9.3 = 10.9$. These two values are plotted on the graph paper and a straight line is drawn through them as shown in Figure 12.2. Make sure that you use a sensible scale when drawing the graphs. It is best, where possible, to use the same scale on both x and y axes.

12.2 The least squares line for a model 1 regression

When we plot two variables on a graph, it is rare to find all the points on a straight line. We are more likely to encounter results similar to those shown in Figure 12.3(a). These data seem to suggest that there is a linear (straight line) relationship between the variables and we might be tempted to draw a line through the cloud of points by hand to give a 'best fit'. This has been done on the scattergraph of Figure 12.3(a), but it is not recommended except in cases where only crude estimates are required (see interpolation in Chapter 7) because it is a subjective procedure and it is most unlikely that another person would draw the line in the same position. Fortunately, there is a formal method which provides a unique 'least squares' line for the graph.

There is a line that passes through the cloud of points, which is positioned so that the total of the deviations of the points from the line is a minimum. These deviations are shown as d values in Figure 12.3(b). Simply summing the deviations will prove unhelpful as some will be positive (i.e. above the line) and others negative (below the line). When the positive and negative deviations are summed, much cancelling out occurs. Therefore the minimum sum of the *squared* deviations is sought to provide the best fit line

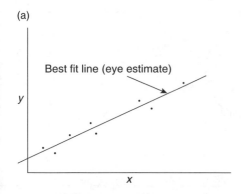

Best fit line (eye estimate)

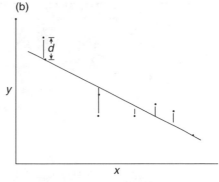

Figure 12.3 (a) A best fit line drawn by eye through a scatter of points;
(b) deviations (d) from a least squares regression line.

(compare the standard deviation calculation, Chapter 3). Using differential calculus, it has been shown that the gradient (b) of the least squares line is given by

$$b = \frac{\Sigma xy - (\Sigma x \Sigma y / n)}{\Sigma x^2 - (\Sigma x)^2 / n}$$

In order to draw the regression line we also need to find a value of the intercept (a). To do this, use is made of the fact the the mean of x and the mean of y lie at a single point on the regression line. This means that

$$\bar{y} = a + b\bar{x}$$

and by rearranging this equation we find that

$$a = \bar{y} - b\bar{x}$$

BOX 12.1: MODEL 1 REGRESSION LINE CALCULATION

The table below shows some of the data obtained by Leffler and Nyholm on the cadmium concentration in the kidney of bank voles (*Clethrionomys glareosa*) as a function of age. This attractive riverside mammal has undergone a serious decline in the British Isles, thought to be connected with water pollution. Voles of known age were used to conform to a model 1 regression.

Kidney cadmium concentration in bank voles

| Age (months) | Cd ($\mu g\ g^{-1}$) |
x	y
2.5	4.6
3.0	4.0
3.5	5.1
4.0	8.1
6.0	10.3

▶

The computations are similar to those used to obtain the product–moment correlation coefficient, but we also require the means of x and y in order to obtain the intercept (a). First, however, a scatterplot is produced. For this example a rough idea of the gradient can be gained as the points lie approximately in a straight line and for a change in y of ten units there is a change in x of about six units. The coefficient b therefore probably lies between 1 and 2. The figure also suggests that the intercept lies within two units of zero.

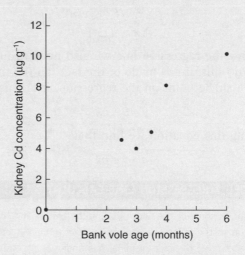

It is important to note that the age of the *Clethrionomys* has been plotted on the x axis as the *independent variable*, as it has been measured without error, with the *dependent* cadmium concentrations plotted on the y axis.

The following values obtain:

$\bar{x} = 3.8$	$\bar{y} = 6.42$
$\sum x = 19$	$\sum y = 32.1$
$\sum x^2 = 79.5$	$\sum y^2 = 234.87$
$\sum xy = 135.55$	$n = 5$

The last term, $\sum xy$, is obtained by multiplying across and summing, i.e. $(2.5 \times 4.6) + (3.0 \times 4.0) +$ etc. If you are using a calculator without a regression function, follow the procedure described for covariance in Chapter 11.

Employing the equation for b above gives

▶

$$b = \frac{135.55 - (19 \times 32.1)/5}{79.5 - (19)^2/5}$$

$$= 13.57/7.3 = 1.859$$

The intercept a is now found:

$$a = 6.42 - (1.859 \times 3.8) = -0.644$$

Hence the regression equation is

$$y = -0.644 + 1.859x$$

12.3 Significance of model 1 regression coefficients

Even though a model 1 regression is appropriate for cases where one variable is measured without error, the error in the dependent variable may be so large that it will not be possible to discern a linear relationship in the scatterplot. Clearly it would be useful to have a statistical test to determine whether there is a significant relationship between the variables – if there is not, there would be little point in continuing the analysis. Product–moment correlation cannot be used as we do not have a bivariate normal distribution. What is required is a test of significance of the regression line itself. This is better understood when it is remembered that when the slope is zero there can be no relationship between x and y. The null hypothesis for the significance test, H_0, states that the *population slope* $\beta = 0$. The alternative hypothesis (two-tailed) states that $\beta \neq 0$ and there is a relationship between the variables.

The significance test can be conducted as a t test or as an analysis of variance as shown below.

12.3.1 t test of the regression coefficient (b)

Here, the t statistic is obtained by dividing the regression line slope (b) by an appropriate standard deviation. This is easier to understand if we consider a whole series of data sets for the same two variables. We can keep sampling the data set, taking several values at a time, to generate a whole series of b values. We can then take the b values as samples in their own right and obtain their standard deviation. Based on the normal distribution, we can go further and generate a value of t by dividing a b value by its standard deviation in a way analogous to a two-sample t test.

BOX 12.2: A t TEST APPLIED TO MODEL 1 REGRESSION COEFFICIENT b

To obtain the standard deviation of b, we do a three-stage calculation. First the residual variance s_{yx}^2 is found. This is calculated from the sums of the squared deviations of both y and x and the sum of products term. Next the standard deviation of the slope (s_b), is obtained by dividing the residual variance by the sum of the squared deviations for x and taking the square root. Finally, t is obtained as the slope (b) divided by s_b and the value is compared with $t_{0.05, n-2}$.

Using the bank vole data., the error term is

$$s_{yx}^2 = SS_y - \frac{(\text{sum of products})^2}{SS_x} \frac{1}{n-2}$$

Several of the components have already been obtained during the regression equation calculation in Box 12.1 and it remains to find the sum of the squared deviations for y (SS_y).

$$SS_x = \Sigma x^2 - (\Sigma x)^2/n = 79.5 - (19)^2/5 = 7.3$$
$$SS_y = \Sigma y^2 - (\Sigma y)^2/n = 234.87 - (32.1)^2/5 = 28.788$$

These values may now be substituted into the equation for the residual variance term give

$$s_{xy}^2 = SS_y - \frac{(\Sigma xy - \Sigma x \Sigma y/n)^2}{SS_x} \frac{1}{n-2}$$

$$= \left[28.788 - \frac{(13.57)^2}{7.3} \right] \frac{1}{3}$$

$$= 3.5627 \times 0.3333 = 1.1875$$

Now calculate s_b as

$$s_b = \sqrt{(s_{error}^2/SS_x)}$$

$$= \sqrt{(1.1875/7.3)}$$

$$= \sqrt{(0.162\ 67)} = 0.403\ 33$$

Finally

$$t = b/s_b$$

$$= 1.859/0.403\ 33 = 4.609$$

The critical value of t with $5 - 2 = 3$ degrees of freedom and $p = 0.05$ (**Table III**) ($t_{0.05,10}$) is **3.182**. Our test value of 4.607 is greater than the critical value, so we reject the null hypothesis that $\beta = 0$ and accept the alternative: that the true (population) regression line is non-zero at the 5% level of significance and that the two variables are related. Note that a one-tailed test should be used if the direction of the population slope parameter (β; positive or negative) is predicted.

Occasionally it may be necessary to compare regression lines with each other to determine whether they come from the same population. For example, in toxin accumulation studies, several species of aquatic animals may be incubated with a range of concentrations of a toxicant. This will yield a series of lines, one for each animal species. Significance tests for either a or b are available. Where only two regressions are being compared a simple t test is available, which is obtained by dividing the difference between the two values of b by the appropriate standard error of the difference. When several regressions are to be compared, a more generalised and involved technique known as the analysis of covariance (ANCOVA) is available (see Sokal and Rohlf, 1995; Underwood, 1997).

12.4 Confidence intervals of the regression coefficients

If it is assumed that the slopes b of a regression are normally distributed then it is possible to fit confidence intervals to the slope. This can be done by multiplying the appropriate value of t_{n-2} by s_b.

For the bank vole data

$$95\% \text{ confidence interval of } b = \pm t_{0.05,\ 3} \times s_b$$

$$= \textbf{3.182} \times 0.403\ 33$$

$$= 1.859 \pm 1.283$$

or

$$b = 3.141 \text{ to } 0.576$$

This means that the true (population) value of the slope β lies within these limits with 95% confidence.

12.4.1 Confidence interval for the coefficient a (the intercept)

It is sometimes useful to calculate confidence intervals to a to decide whether it is significantly different from zero or some other value. The calculation is made using the equation

$$a \pm t_{\alpha,\ n-2} \sqrt{\left[s_{yx}^2 \left(\frac{1}{n} + \frac{\bar{x}^2}{\mathrm{SS}_x} \right) \right]}$$

12.5 Model 1 regression and the analysis of variance

An analysis of variance can be performed by dividing the total sum of the squared deviations of a regression into two components. These are a component measuring how much of the variation among the y values can be explained by the fitted regression alone, and a component measuring the size of the deviations from this line. The second component is of particular importance because, if it is small, the line will be a good fit. It is best calculated by first obtaining the sum of the squared deviations for the regression and then subtracting it from the total sum of the squared deviations.

Using the example of the bank vole data we have

$$\text{total sum of squared deviations for } y, \ \mathrm{SS}_{\text{total}} = \Sigma y^2 - (\Sigma y)^2/n$$

$$= 234.87 - (32.1)^2/5 = 28.79$$

The sum of squares for regression is given as

$$\mathrm{SS}_{\text{regression}} = \frac{(\text{sum of products})^2}{\mathrm{SS}_x}$$

For the example

$$(13.57)^2/7.3 = 25.23$$

Finally, we obtain the error term (often called the *residual* or *unexplained* sum of squared deviations) as

$$SS_e = SS_{total} - SS_{regression}$$

$$= 28.79 - 25.23 = 3.563$$

Now the ANOVA table can be completed:

Source of variation	df	SS	MS	F
Regression	1	25.23	25.23	21.2
Error	3 ($n-2$)	3.563	1.188	
Total	4 ($n-1$)	28.79		

Consulting the critical values of F with one and three degrees of freedom (**Table VIII**) shows that the result is significant at the 1% level. Note that the error mean square value of 1.1875 is identical to that obtained in the t test of the slope b ($= s^2_{error}$). In fact, this quantity is equal to $\Sigma(y_{observed} - y_{predicted})^2$ or Σe^2.

12.6 The coefficient of determination

The coefficient of determination (r^2) is often quoted in regression ANOVA because it is a measure of the amount of variability in one variable that is accounted for by the variability in the other. It is equal to the square of the product–moment correlation coefficient and is also the ratio of $SS_{regression}/SS_{total}$. The closer the points to the regression line, the greater is the proportion of the explained (regression) variation to the unexplained (error) variation.

12.7 Origin forcing

There may be occasions when we have good reason to believe that the population regression line will pass through the origin of the graph. With the bank vole data it would impossible to measure kidney cadmium before the kidney had differentiated in the foetus. Thus there must be a true zero point, and negative x values (i.e. negative times) would be meaningless. There may be other cases where we have reason to believe that the population intercept α is zero, and we can test the hypothesis by obtaining the confidence interval for a as outlined above and determining whether it overlaps zero. This test can also be used to determine linearity. If the value of a is distant from the origin, the relationship may well be non-linear.

BOX 12.3: REGRESSION ORIGIN FORCING

The bank vole data can be used to demonstrate origin forcing because the coefficient a is close to zero. There is good reason to expect that $\alpha = 0$, since a vole at age 0 will contain zero cadmium! In ambiguous situations, obtain the 95% confidence interval for α. If the value 0 is included then you can be reasonably certain that origin forcing is appropriate.

The new equation for b is

$$b = \Sigma xy / \Sigma x^2$$

These values have already been calculated :

$$\Sigma xy = 135.55 \qquad \Sigma x^2 = 79.5$$

b with origin forcing becomes

$$b = 135.55/79.5 = 1.705$$

giving the new regression equation

$$y - 1.705x$$

12.8 The prediction interval and confidence interval for estimates of *y*

One of the most important applications of regression is to find estimates of *y* values for given *x* values. With the bank vole data a significant linear relationship between kidney cadmium concentration and age was found. A researcher could use the regression equation

$$y = -0.644 + 1.859x$$

to obtain a value of the kidney cadmium level at age 5 months. When $x = 5.0$,

$$y = -0.644 + 1.859 \times 5 = 8.651 \ \mu g \ g^{-1}$$

This value has been obtained from the line of best fit, but we already know that it can only be an estimate; for if the entire study were repeated, another value of *b* would be obtained yielding a different value of *y*. Therefore, it would be useful to obtain some measure of the error in *y*. This leads to a consideration of prediction and confidence intervals for *y*.

Prediction and confidence intervals are both interval estimates, but they measure different quantities. A confidence interval will tell us the range within which a single *y* value will lie with 95% probability. It is used when we have a regression equation for an experiment and we wish to estimate a *y* value from a given *x* value. A prediction interval will give the range in which the mean of *all y* values (from a given *x* value) will occur. It is used if further experiments are performed and it is necessary to predict a *y* value from these other experiments using the regression equation from the

BOX 12.4: PREDICTION INTERVAL CALCULATION

Using the bank vole data, find the prediction interval to a $\hat{y}$ value corresponding to $x = 5$ months.

Using the regression equation above, the value of $\hat{y}$ has already been obtained as $\hat{y} = 8.651$ μg g^{-1} Cd. The following quantities have already been calculated:

$SS_x = 7.3$ $SS_y = 28.788$
$\bar{x} = 3.8$ sum of products $= 13.57$
$n = 5$ $x_i = 5$

The variance from the regression line (the residual variance) becomes

$$s^2_{yx} = \left[28.788 - \frac{(13.57)^2}{7.3} \right] \frac{1}{3}$$

$$= (28.788 - 25.225) \times 0.333\,33$$

$$= 1.1875$$

The prediction interval becomes

$$t \times \sqrt{\left\{ 1.1875 \left[1 + \frac{1}{5} + \frac{(5 - 3.8)^2}{7.3} \right] \right\}}$$

$$= t \times \sqrt{(1.1875 \times 1.3973)}$$

$$= t \times \sqrt{(1.6593)}$$

$$= t \times 1.288$$

For 95% confidence intervals, the critical value of t is the two-tailed $p = 0.05$ value with $n - 2$ degrees of freedom. Consulting **Table III** gives $t_{0.05,\,3} = \textbf{3.182}$. Thus the 95% prediction interval is $\hat{y} \pm \textbf{3.182} \times 1.288$ or

$$8.651 \pm 4.098 \text{ (or } 4.553 - 12.75) \text{ μg g}^{-1}$$

original experiment. Prediction intervals are always wider than confidence intervals and they are calculated in a similar way.

The formula used to obtain a prediction interval is

$$\hat{y} \pm t \sqrt{\left\{ s^2_{xy} \left[1 + \frac{1}{n} + \frac{(x_i - \bar{x})^2}{SS_x} \right] \right\}}$$

where x_i is the x value corresponding to $\hat{y}$ (read as y hat).

12.9 The confidence interval for y

Occasionally it is useful to obtain confidence limits to the mean of all y values corresponding to x_i, rather than a single value as calculated above. The calculation is almost identical to the prediction interval calculation, differing only in the omission of unity in the square root expression:

BOX 12.5: CONFIDENCE INTERVAL CALCULATION

For the bank vole data, the calculations are thus very similar to those made in the prediction interval

$$t \times \sqrt{\left\{1.1865\left[\frac{1}{5} + \frac{(5 - 3.8)^2}{7.3}\right]\right\}}$$

$$= t \times \sqrt{(1.1865 \times 0.3973)}$$

$$= t \times \sqrt{(0.4713)}$$

$$= t \times 0.6865$$

The 95% confidence interval becomes

$$\hat{y} \pm \mathbf{3.182} \times 0.6865$$

$$= 8.65 \pm 2.185 \text{ (or } 6.465 \text{ to } 10.835) \ \mu g \ g^{-1} \ Cd$$

The confidence interval is therefore narrower than the prediction interval. The relationship between these two interval estimates and the regression line is shown graphically in the Minitab example later.

$$\hat{y} \pm t \times \sqrt{\left\{\left\{s_{yx}^2\left[\frac{1}{n} + \frac{(x_i - \bar{x})^2}{SS_x}\right]\right\}\right\}}$$

12.10 Model 1 regression for cases where there are several y values for each x value

It is common in experiments to replicate samples taken at discrete time intervals, to provide more reliable information. In the example above, only one of the y values was taken for each of the x values. If we look at the complete data set of Leffler and Nyholm (Table 12.1), you can see that several y values were obtained for some x values. In such cases the regression coefficients are obtained in the normal way, but the regression statistics are obtained using ANOVA.

When these data are graphed a linear relationship is still apparent and the regression of age on kidney cadmium remains significant. However, if we wish to perform significance tests or set confidence levels, the formulae given above must be modified. Also, note the outlier (4.0, 15.2), which will have a considerable influence on the calculations.

Confidence intervals are now calculated using

$$\hat{y} \pm t \sqrt{\left\{\left\{s_{yx}^2\left[\frac{1}{n} + \frac{(x_i - \bar{x})^2}{SS_x}\right] + MS_{within}\right\}\right\}}$$

Table 12.1 Complete bank vole data showing several y values for some of the x values

Age (months) x	Kidney Cd($\mu g\ g^{-1}$) y
2.5	3.3
2.5	3.6
2.5	3.9
2.5	4.6
2.5	6.7
3.0	4.0
3.5	5.1
4.0	7.2
4.0	8.1
4.0	8.35
4.0	15.2
6.0	10.3

where MS_{within}, the within samples mean square (i.e. variance), is obtained from two sums of squares, as in the analysis of variance

$$SS_{within} = SS_{total} - \Sigma[(\Sigma y^2/n_i) - (\Sigma)^2/n_i]$$

with degrees of freedom $(n-1)(i-1)$, where i is the number of groups (i.e. the different x values) and n_i is the number of y values in the ith group. For the bank vole data, with $x = 2.5$ months, $n_i = 5$, $i = 5$ and $n = 12$.

For prediction intervals the equation used is the same as that described above, but to test the significance of the regression an analysis of variance is required.

12.11 Model 2 regression

If a regression line is required for bivariately normal distributions, a model 2 regression analysis will be required. For a model 1 regression, it is assumed that one variable, x, is fixed and has negligible associated error, while the dependent variable y is subject to random error with a normal distribution. In model 2 regression, there is no independent variable so there is a cloud of variability surrounding each point on the graph, as shown in Figure 12.4. It is better to assign the two variables x_1 and x_2, rather than y and x.

The regression coefficient b' (b prime) is most simply obtained as

$$b' = s_{x_1}/s_{x_2}$$

In other words, the sample standard deviation of one of the variables divided by the sample standard deviation of the other. If one variable is to be estimated from the other, which is usually the case, then x_1 (equivalent to y in model 1 regression) is made the estimated variable. The slope may be either

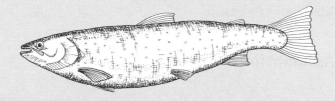

Figure 12.4 Direction of variation assumed for model 1 and model 2 regression.

BOX 12.6: MODEL 2 REGRESSION

This method will be required most frequently in field observations where there is usually little chance of controlling either of the variables. A sample of 4-year-old charr (*Salvelinus alpinus*) was analysed for total PCB (polychlorinated biphenyl) concentration per gramme of fish and compared with the kidney biomass.

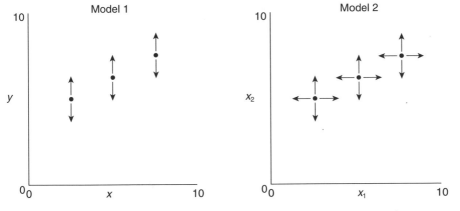

Charr kidney biomass and PCB data

Kidney biomass (g dry wt) x_1	PCB concentration ($\mu g \ g^{-1}$) x_2
11.25	4.6
10.37	3.8
8.82	6.5
8.15	7.05
8.5	9.85
9.0	4.55
10.15	7.2
11.81	2.0
10.18	5.65
9.16	7.15
$\bar{x}_1 = 9.739$	$\bar{x}_2 = 5.835$
$s_{x_1} = 1.2073$	$s_{x_2} = 2.2032$
$n = 10$	

$$b' = s_{x_1}/s_{x_2} = 1.2073/2.2032 = -0.548$$

By inspection of the scattergraph b' is deemed negative.

$$a' = \bar{x}_1 - b'\bar{x}_2$$

$$= 9.739 - (-0.548 \times 5.835) = 12.94$$

The regression equation becomes

$$x_1 = 12.94 - 0.548x_2$$

For example, if the kidney mass is 9.5 g, the PCB concentration is estimated as

$$x_1 = 12.94 - 0.548 \times 9.5 = 7.73 \ \mu g \ g^{-1}$$

positive or negative and this can usually be decided from the graph. Alternatively, the sign of the product–moment correlation coefficient can be used. The intercept a' is obtained as in model 1 regression, namely as $\bar{x}_1 - b'\bar{x}_2$.

Significance testing is not as straightforward as in model 1 regression and requires an analysis of variance.

12.12 Assumptions

As a technique involving the normal distribution, regression analysis may only be applied subject to certain conditions. One way of testing for normality is to plot the **residuals**.

A residual is the quantity

$$e = y - \hat{y}$$

$$= y - (a + bx)$$

Its relationship to the regression line can be seen in Figure 12.5. The residual e is an estimate of a quantity ε, which for regression must be homoscedastic, normally distributed and with zero mean. These properties are readily checked by plotting the residuals. Example plots are shown in Figure 12.5.

In Figure 12.5(a), some residuals have been plotted against x. The residuals cluster around the value zero with approximately the same number above and below the line and there is an even spread along the regression line. This plot suggests both normality and homoscedasticity. In Figure 12.5(b) there is significant departure from linearity as might occur in a curvilinear relationship and plot 12.5(c) demonstrates heteroscedasticiy where the variance increases from left to right. Finally, in Figure 12.5(d) a systematic trend

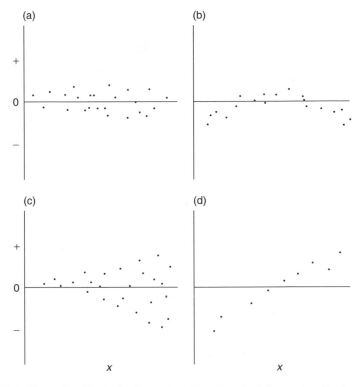

Figure 12.5 Plots of residuals for four regressions (see text for explanation).

is apparent. This suggests some form of dependence, which again violates the assumptions.

Regression analysis using Minitab
A model 1 regression can be performed by inputting the two variables into columns 1 and 2 of a worksheet:

C1	C2
13	30
17	32
7	24
24	35
28	39
9	25
6	23
20	36
24	38
14	29
18	35
10	29

On the menu bar:

Stat > Regression > Regression

Type C2 in the 'Response' box and C1 in the 'Predictors' box. To examine the residuals, click on 'Storage' and click the boxes labelled 'fits' and 'residual' so that ticks appear in the boxes; then click 'OK' twice.

```
The regression equation is
C2 = 19.7 + 0.730 C1
Predictor           Coef        StDev          T          P
Constant          19.687       1.104       17.84      0.000
C1                0.73032      0.06390      11.43      0.000
S = 1.528    R-Sq = 92.9%    R-Sq (adj) = 92.2%
Analysis of Variance
Source              DF          SS           MS         F         P
Regression           1        304.91       304.91    130.63    0.000
Residual Error      10         23.34         2.33
Total               11        328.25
```

The regression equation is given in the form $C2 = a + bC1$; thus C1 represents x and C2 represents y. Below the regression equation are several tables. The first beginning with 'Predictor' gives the values of a under 'constant coefficient' and b under 'C1 coefficient'. Their standard deviations and t values are also shown; the standard deviation for C1 being s_b, as described above. The extreme right-hand column shows their probabilities under H_0. Both are seen to be low (they have been rounded to zero) showing them to be highly significant. In the case of b (= 'C1 coefficient') this means that the regression slope is significantly different from zero. In the case of the intercept a ('constant coefficient') it means that its value is also significantly different from zero.

Next there is a row of values beginning with s. This is the standard deviation from the regression line, which is followed by r^2, which is the coefficient of determination. The quantity r^2(adj) is a population estimate of the coefficient rather than the sample estimate.

Finally, an analysis of variance is provided for the regression, which follows the form given in the text above.

Predictions with Minitab
To obtain a predicted value from the equation and its confidence interval perform the following:

Stat > Regression > Regression

Type C2 in the 'Response' box and C1 in the 'Predictors' box and then click 'Options'. Type 17.5 in the box after 'Prediction intervals for new observations'. Click 'OK' twice.

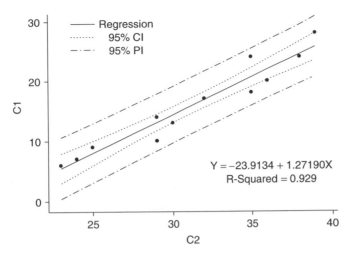

Figure 12.6 Regression line, prediction and confidence intervals.

The output provides the predicted value of *y* for the given value of *x* (17.5).

```
Predicted Values
  Fit        StDev Fit         95.0% CI             95.0% PI
32.467        0.454        (31.456, 33.478)     (28.916, 36.018)
```

The 'Fit' is the value of $\hat{y}$ for the *x* value 17.5. The 95% confidence and prediction intervals are given by default. If you want to change the confidence level (e.g. to 90%) this can be done by typing 90 in the 'Confidence level' box.

Plotting the regression line and confidence intervals with Minitab

Stat > Regression > Fitted Line Plot

In the 'Response' box put C2 and in the 'Predictor' box put C1.

A linear regression model is required, and under 'Display options' click on 'Display confidence bands' and 'Display predictor bands' if you wish to display both of these statistics graphically. Then click 'OK' twice.

The regression line and confidence bands will be displayed with the regression equation, as shown below. Note that the confidence and predictor bands curve in towards the regression line, their closest approach being in the region of $\bar{x}$ on the regression line (Figure 12.6).

Plotting the residuals with Minitab

This can be performed using the same worksheet. In the first exercise, the residuals and fits were stored in C3 and C4 by default. Perform the following:

Stat > Regression > Residual Plots

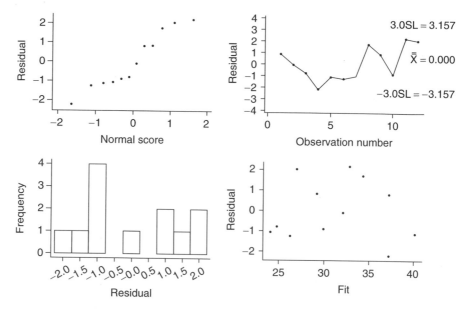

Figure 12.7 Plots of residuals (Minitab).

In the dialog box, enter C4 in the 'Residuals' box and enter C3 in the 'Fits' box; then click 'OK'. The output is shown in Figure 12.7 as a series of graphs displaying the residuals and fits.

In this example, the residuals do not fit the normal distribution well. The top left graph shows a normality plot similar to that used to test data with probability paper. The points are not spread along a straight line and suggest kurtosis or bimodality. The plot of the residuals as a frequency histogram (lower left) supports this. The plot in the upper right-hand graph illustrates how the residuals change along the regression line. They appear to be homoscedastic but, with few data points, caution should be exercised. If these were experimental results, the regression analysis would need to be used with care, and ideally the experiment should be repeated.

12.13 Non-parametric regression

Many scattergraphs will indicate a non-linear relationship between two variables. In some cases the relationship can be made linear by a simple mathematical transformation (see Chapter 13). An alternative is to perform a non-parametric regression. An example is Kendall's robust line-fit method (Kendall and Gibbons, 1990), where gradients are calculated for every pair of x values to provide a non-parametric value of b and a regression equation.

- There are two regression models, model 1 and model 2, suitable for describing and testing significant relationships between two variables.

- Model 1 regression is used when one of the variables is fixed so that it is measured with negligible error.

- In model 1 regression the fixed variable is called the independent variable x and the dependent variable is designated y.

- The regression equation is written $y = a + bx$, where a and b are the regression coefficients.

- The regression equation is used to draw a best fit straight line on a scattergraph.

- Model 2 regression is used where neither variable is fixed and both are measured with error.

- Significance tests are available to determine whether the population regression coefficients α and β differ from zero and establish if there is a significant relationship between variables x and y.

Exercises

12.1 The data below show air temperature in a mine as a function of depth below the surface. Obtain the model 1 regression equation and estimate the temperature at a depth of 1000 m.

Depth below ground (m)	Air temperature (°C)
0	10.6
140	11.6
300	13.3
170	13.8
310	15.0
340	17.0
470	17.2
460	19.3
550	20.6
840	22.1
690	22.6
590	23.6
820	25.6
1020	26.9
1150	30.2
970	30.6

1340	31.1
1130	33.0
1290	33.7
1410	33.9
1500	36.0 ($n = 21$)

12.2 Measurements of water temperature were made in an alpine stream and the altitude was recorded between 150 and 3750 m above sea level. Graph the altitude against water temperature and obtain the model 1 regression equation. Estimate the water temperature at 3000 m altitude and the 95% confidence limits of this estimate.

Altitude (m)	Water temperature (°C)
150	19.5
400	13.0
960	12.3
970	11.5
1200	9.8
1360	11.8
1370	10.2
1570	8.8
1590	9.0
1700	11.7
1920	6.2
2020	5.8
2300	7.0
2510	5.0
2800	3.3
3400	3.8
3600	1.2
3750	2.6

12.3 The data below are of average wind speeds taken over a one year period with site altitude for an area of moorland in western England. Obtain the model 1 regression equation for these data. A wind turbine is to be built on the moorland. What is the lowest altitude at which the turbine can be constructed if there is a minimum average wind speed requirement of 7 km h^{-1}? Assume that the turbine blades will operate 25 m above the ground level.

Altitude (m)	Average monthly wind speed (km h^{-1})
177	4.6
181	7.7
92	4.6
240	10.8
86	24.0
83	8.1
45	2.0

265	5.3	
115	7.2	
275	8.3	
115	2.6	
44	4.3	
210	10.6	
35	0.9	
155	2.5	
55	1.4	
238	7.0	$(n = 17)$

12.4 The following data show the accumulation of copper in the shells of the barnacle *Elminius* as a function of time in a toxicity experiment. Plot the data, obtain the model 1 regression equation and perform a t test for the significance of the regression slope β.

Time (days)	Copper accumulated in shell (ppm)	
1.0	4.2	
2.0	8.1	
3.0	4.8	
5.0	8.5	
10.0	18.5	
15.0	24.6	
20.0	40.8	
25.0	46.5	
30.0	36.4	
40.0	73.8	
50.0	81.2	
60.0	87.2	$(n = 12)$

12.5 The following data were obtained for liver mercury concentration ($\mu g\ g^{-1}$) in the kestrel (*Falco tinnunculus*) in relation to body weight. Find the model 1 regression equation with and without origin forcing.

Body weight (kg)	Liver Hg ($\mu g\ g^{-1}$)	
0.121	2.60	
0.205	4.00	
0.322	3.18	
0.305	4.40	
0.395	5.48	
0.535	5.08	
0.538	3.42	
0.566	5.95	
0.602	4.22	
0.681	5.61	
0.776	8.40	
0.835	4.96	
0.938	6.71	$(n = 13)$

CHAPTER THIRTEEN

An introduction to modelling

Models take many forms. Some are literally scaled-down versions of large structures, such as the aeroplanes used in wind tunnel experiments, but most consist of mathematical expressions describing as simply as possible the relationships between the variables of an often complex system and are essentially qualitative abstractions of reality. Some models have been developed by statisticians to obtain reliable information from experimental procedures such as toxicity testing. Other models do not have an explicit statistical element and yet provide useful predictions, such as the growth rate of a population. The key components of a mathematical model are simple and logical relationships between the variables of interest, but care must be taken not to oversimplify a model. By investigating the behaviour of the environment under natural and laboratory conditions it is possible to formulate relationships into a model, which can be 'validated' by applying the model to systems whose behaviour is already known. As an example, we may have a model that predicts the range of particle fallout from a chimney at a given distance. By measuring the chimney height, wind speed and the particle size we might make predictions on fallout using a model. To validate the model, it would be necessary to collect fallout data from a range of chimneys and compare them with the model predictions. If the model provided a reliable estimate of the fallout, then it is said to have been validated and can then be applied to similar chimneys where fallout data are lacking.

Environmental scientists need to understand how processes are modelled and how the models can be tested. Predictive models are often used in studies of resource management, pollution control and environmental impact assessments. They are usually available as computer packages designed for particular situations, though the design and formulation of a simple environmental model is certainly within the grasp of a competent undergraduate student. Another important use of models is to optimise environmental decision making. In this case models can provide a solution to the most cost-effective way of dealing with an environmental problem. Statisticians use a narrower definition of the term *model*, which refers to the mathematical relationship describing a population in terms of its frequency. Well-known models are the normal, binomial and Poisson distributions.

This chapter gives an introduction to several modelling applications, drawing upon some of the statistical techniques learnt in the previous chapters.

13.1 Deterministic models

Mathematical models fall into a number of categories. The simplest models provide an equation relating the value of a single dependent variable to one or more independent variables. For example, a model can be used to estimate the age of birch trees (*Betula* spp.) from their girth (diameter)

$$\text{age (yr)} = 0.636 \text{ girth (cm)}$$

You can see that the value of 'age' is a simple linear function of 'girth' and is identical to a type of regression analysis described in Chapter 12 with the intercept $a = 0$. Such models, which return a single value of the variable of interest but tell us nothing about the variability of the result, are called *deterministic models*. They are the simplest to construct and validate. Providing just a single estimate for a given value of the dependent variable, however, may not be much use if the independent variables could not be reliably measured, as we could then have little confidence in the prediction. This is worrying if we are basing, say, management policies or costs on the results of the model. An improvement can be made by adding a random component e to the model. In doing so, it becomes possible to provide an estimate of the amount of variation we can expect in our predicted values. These are called *stochastic models*. Such models, while providing a more realistic description of nature, are much more difficult to construct.

Models are also classified according to whether or not they include time. If time is not a component of a model, the model is termed a *static* or *instantaneous model*. This does not mean to say that the process being modelled does not change with time, but the process is being evaluated at a particular point in time. If time is a component of the model, then the model is termed a *dynamic model*. These models can be highly complex with dozens of variables described by hundreds of equations. The more complex models can only be run on a computer, as it would take too long to do the calculations by hand. Very large numbers of calculations may be required as some of these models output values for small units of time. A climate model, which is used to predict conditions over several days or weeks, will require 'start values' for all the variables, and to output the results minute-by-minute requires a large amount of computing time.

Models can also be classified according to the number of processes being predicted. A simple model might predict the concentration of oxygen down a watercourse as a function of its distance below a sewage outfall (single process), while another may predict the levels of phosphate, nitrate and ammonia in a lake as a function of time (multi-process). Some examples of simple deterministic and dynamic models are given below to illustrate how the models are constructed and how they are used.

13.2 A deterministic model employing log–log transformations

This example shows how a model can be used to assess lake trophic status using a very simple measurement. It involves the use of a Secchi disc, which is lowered into a lake from the side of a boat. The disc is painted with alternate black and white quadrants, and the depth of water at which the disc disappears from view is recorded. This depth is called the Secchi depth. Robert Carlson (1977) measured the Secchi depths in a large number of lakes together with the chlorophyll-a concentration in the water. The amount of chlorophyll-a is a good measure of phytoplankton biomass. In lakes a large biomass is associated with nutrient enrichment – there is a relationship between chlorophyll-a and trophic status (Table 13.1). Carlson wanted to know if there was a relationship between the Secchi depth and the concentration of chlorophyll-a in the water.

Figure 13.1 is a scattergraph of the data, which shows a non-linear relationship between the two variables. However, by taking the natural logarithm (see Appendix 3) of the Secchi depth and plotting it against the natural logarithm of the chlorophyll-a measurements, a significant linear relationship

Table 13.1 Summer chlorophyll-a value as an indicator of lake trophic status (mg m^{-3})

Oligotrophic	Mesotrophic	Eutrophic
0–2.8	2.8–8.7	>8.7

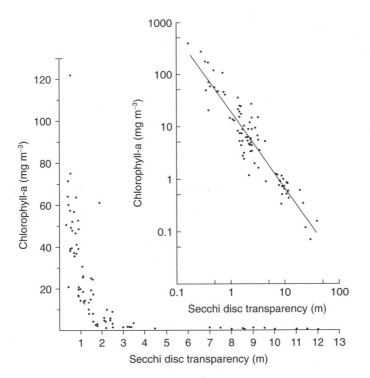

Figure 13.1 Relationship between Secchi disc transparency and near-surface concentrations of chlorophyll-a. Inset shows the log–log transformation of the same data. Redrawn from Carlson (1977) *Liminology and oceanography*.

becomes apparent (Figure 13.1, inset). A mathematical transformation followed by regresssion analysis enabled the relationship to be expressed in simple terms.

Linear regression of the transformed data yielded

$$\ln SD = 2.04 - 0.68 \ln Chl\text{-}a$$

where SD is Secchi depth measured in metres and Chl-a is chlorophyll-a measured in mg m^{-3}. This equation provides a simple relationship between the variables and allows lake trophic status to be determined from the data in Table 13.1.

As an example, suppose a lake, suspected of being enriched as a consequence of agricultural runoff, has a summertime Secchi depth of 2.15 m. Substituting this value in the model

$$\ln 2.15 = 2.04 - 0.68 \ln Chl\text{-}a$$

$$0.7655 = 2.04 - 0.68 \ln Chl\text{-}a$$

$$0.68 \ln Chl\text{-}a = 2.04 - 0.7655 = 1.2745$$

Thus

$$Chl\text{-}a = 6.52 \text{ mg m}^{-3}$$

With reference to Table 13.1 we can see that the lake would be classified as mesotrophic. If other lakes in the catchment were found to be oligotrophic (nutrient-poor), there could be reason for concern.

13.3 Predictive model for biomass of nuisance algae in lakes

The blue-green algae (cyanobacteria), such as *Anabaena*, often form dense waterblooms in fertile lakes rendering the water mildly toxic and making the water difficult to filter at water treatment plants. This is a particular problem in areas of high population density where the water has considerable amenity value and where there is a high demand for potable water. Smith (1985) concluded that the biomass of cyanobacteria was most probably related to the nutrient status of the water (the P and N content, measured in mg m^{-3}) and the lake depth (Z in metres). Information on total N, total P and Z were obtained, together with the value of total algal cell volume, from five sites visited on many occasions. Correlations were then obtained between these combinations of variables. Regression analysis was performed on the selected log-transformed variables, as this was found to stabilise the variance and reduce the heteroscedasticity. Note that the regression analysis was undertaken using two independent variables (P, Z) rather than the usual one, a procedure known as *multiple regression*, which is available in most statistics packages. The following model was obtained

$$\log BG = 0.596 \log P - 0.963 \log Z - 0.142$$

If, for example, a lake had a P value of 870 mg m^{-3} and a depth of 11.8 m, the predicted cyanobacterium biomass is

$$\log BG = (0.596 \times 2.939) - (0.963 \times 1.072) - 0.142$$

$$= 1.752 - 1.032 - 0.142 = 0.578$$

Then

$$BG = 10^{0.578} = 3.78 \text{ g m}^{-3}$$

This is a fairly high biomass, which could cause problems if the lake is to be used for watersports. The application of these and similar models can be used to assess the impact of nutrients on lakes, providing useful lake management tools.

It is possible to see how a stochastic model might be developed in the above examples incorporating statistical information from the original data. For each regression, confidence intervals could be calculated for the predicted y values and used to provide a range of output values for the given input values. The above models are *static*, as they provide information for a particular instant in time. They could be converted into dynamic models if the relationship between phytoplankton biomass and nutrient levels with the time of year could be described mathematically. In practice this is very difficult

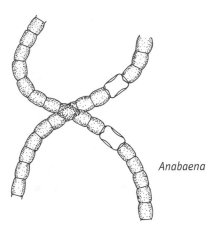

Anabaena

because a large number of other factors affect these variables and many of the interrelationships may be inadequately described.

13.4 A simple dynamic model: the growth of mammals using the Gompertz equation

The Gompertz equation belongs to the Richards family of simple dynamic models. It provides an estimate of an animal's mass as a function of time and is described by

$$W(t) = W_{max} \exp\{-\exp[-k(t - I)]\}$$

where $W(t)$ is the animal's mass in grammes at time t, W_{max} is the asymptotic (maximum) mass in grammes, k is a growth rate constant (day^{-1}) and I is the inflection point (days).

The model has the property that the relative growth rate of the animal declines linearly with log(mass) and this provides a good description of the growth processes occurring in most mammals. The resulting growth curve is sigmoidal (S-shaped) or convex, showing a short lag and then rapid (exponential) growth, followed by a steady decline to a growth rate of zero (Figure 13.2). The model can be fitted to the growth of any animal for which suitable measurements are available. Each species requires its own set of parameters (the quantities W_{max}, k and I) and these can be obtained using a suitable regression analysis on an experimental set of animals. Note that, in the modelling literature, the term *parameter* is used in a more general sense than it is by statisticians and refers to any fixed value in the model equation. Thus the 'parameters' of the Gompertz model, which vary from species to species, are estimated from animal *samples* and are really *statistics*. In a study of over 300 species of mammals, Zullinger *et al.* (1984) obtained good fits for growth using the Gompertz model; an example of its use is given below.

For bushbuck (*Tragelaphus scriptus*), the following parameters can be used for the Gompertz model: $W_{max} = 28\,500$ g; $k = 0.0044$ day^{-1}; $I = 150$ days. We

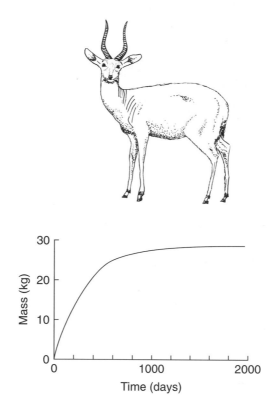

Figure 13.2 The Gompertz model fitted to the growth of bushbuck.

shall estimate the mass of a bushbuck of age (t) 50 days by substituting these parameters into the Gompertz equation.

$$W(t) = 28\,500 \exp\{-\exp[-0.0044(50 - 150)]\}$$

Since the model includes two exponents, the calculation is performed in stages.
   For the higher exponent

$$\exp[-0.0044(50 - 150)] = \exp[-0.0044 \times (-100)]$$
$$= \exp[-(-0.44)]$$
$$= \exp(0.44) = 1.553$$

giving for the lower exponent

$$\exp(-1.553) = 0.2116$$

and

$$W(t) = 28\,500 \times 0.2116 = 6031 \text{ g } (6.03 \text{ kg})$$

This is an example of a very simple dynamic model. Models which include several variables in addition to time need an equivalent number of equations to describe their interrelationships. This leads to the use of *iterative* procedures, where the value of each variable changes after each time step. A discontinuous, stepped curve results, describing the change of the variable over time. In the case of the Gompertz model, where a change in a single variable (mass) is calculated, recalculation of variables after each time step is not required and the resulting time series is a smooth curve.

Key notes

- Models are described by mathematical expressions and are used to predict events or conditions in the environment.

- By giving the mathematical expressions test values, models can be used to predict a range of outcomes, making them useful for environmental impact assessments.

- Static models provide predictions at one instant in time.

- Dynamic models provide predictions as a time series.

- Deterministic models output a discrete prediction and provide no information concerning the error of the prediction.

- Stochastic models can provide information which allows the error of a prediction to be estimated.

CHAPTER FOURTEEN

An introduction to toxicity testing

Controlling the quality of the food we eat and the water we drink involves extensive testing of the potentially dangerous chemicals which could enter the food chain. To understand how the biological components of ecosystems respond to toxic pollutants is also important so that the impacts of these pollutants can be assessed. Several standard methods of testing chemicals for safety are routinely undertaken in laboratories. Most of these tests involve exposing animals, plants or cell cultures for a short period of time to known levels of a potential toxicant. Toxicity may be *acute*, when the survival of the organism is a few hours or days, or *chronic*, where the effects may take many months or years to develop. Acute toxicity tests are undertaken under laboratory conditions over a period of hours or a few days. Usually, toxicity tests expose a test organism to a range of concentrations of toxicant over a period of two–ten days. Suitable 'control' samples, where the toxicant is absent, must be included.

14.1 Dose–response curves and probits

The data in Table 14.1 are the results of a toxicity test using roach (*Rutilus rutilus*) exposed to a solution of the insecticide dieldrin at a concentration of 30 µg 1^{-1} over 16 days. Twenty fish were exposed to the toxicant in a large tank in which the solution was slowly circulated.

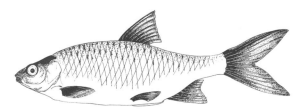

Table 14.1

Time (min)	Cumulative percentage mortality
5960	5
6580	10
7550	20
8440	30
9660	45
11 220	60
13 490	75
14 480	80
16 850	90
22 900	100

The table was prepared by measuring the time it took each fish to die. The cumulative percentage mortality was then calculated. For example, the first fish died after 5960 min of exposure. Since there were 20 fish in the tank, this represents 1/20 or 5% of the total, so five was recorded. After 11 220 min a total of 12 fish were dead, so the figure 12/20 × 100% is recorded. All of the fish had died after 22 900 min. When these data are graphed (Figure 14.1(a)) an S-shaped curve is obtained, but the shape of the S is not symmetrical – the lower bend of the S is much shorter than the upper bend. This form is called a skewed sigmoid curve and is commonly observed in toxicity studies. By transforming the time axis to log(time) Figure 14.1(b) is obtained. This curve looks much more like a symmetric S, the skewness having been largely eliminated.

If you turn back to Figure 1 in Box 7.2 (Chapter 7) you will see the similarity between the curve in Figure 14.1(b) and the curve of the cumulative percentage frequency for a normal distribution. One of the normality testing procedures is to plot cumulative percentage frequency on normal probability paper to see whether a straight line is obtained. In other words, if the curve represents a normal distribution, it is 'linearised' by transferring it to normal probability paper. In the same way, the toxicity curve showing log(time) on the abscissa is also linearised using probability paper. This enables some useful statistical quantities to be determined.

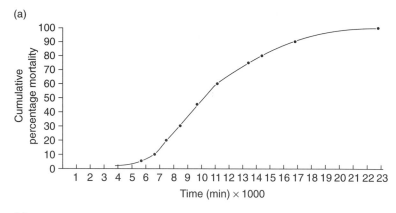

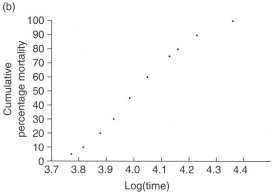

Figure 14.1 (a) Roach toxicity data showing a skewed sigmoid curve; (b) roach cumulative mortality showing the result of the log transformation.

Figure 14.2 shows the result of plotting the toxicity data on probability (probit) paper. Usually a special log-probability paper is used, which avoids the calculation of logarithms. The paper is first oriented so that the figures 1/0.01 appear at the bottom left-hand corner. The cumulative percentage mortality is now plotted along the probability axis (abscissa), with time along the ordinate, which provides a logarithmic scale. Normally this axis permits values in the range 1–1000 to be plotted. Alternatively, a range of 10–10 000 or 100–100 000 can be used. For our data, we need to plot in the range 100–100 000, since the longest time is 22 900 min. Note that is it not possible to plot zero time on log-probability paper, nor 100% mortality.

The best-fit line is drawn using a transparent ruler through the points, paying most attention to the points lying between 40% and 80% cumulative mortality.

14.2 The LT50

An important statistic in toxicity studies is the median survival time or LT50. This time is an important characteristic for an organism's response to a

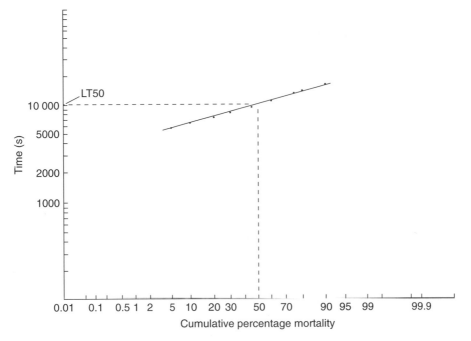

Figure 14.2 Probit of roach toxicity data showing the estimation of the LT50.

toxicant. The LT50 should be a fixed value for a particular species' response to a toxicant, although it will vary to some extent with temperature and composition of the solution. If the temperature and composition are standardised, the LT50 can be used to compare toxicities of compounds and to set legal standards and permitted levels in the environment. These are achieved using 'lethal threshold concentrations'.

The LT50 can be estimated from a probit plot by producing the 50% cumulative mortality to the best-fit line. The time can then be read off the log axis. For the example above, the LT50 is estimated as 10 025 min.

Bliss (1937) and others recommend a slight modification of toxicity data prior to plotting to account for the small number of measurements characteristic of these studies. The modification helps to account for animals which show extreme responses and which may not have been sampled in the investigation. It is made by reducing the observed percentage mortality by one-half of the percentage value contributed by a single animal in the sample. This reduction is applied to all the mortality figures and is equivalent to subtracting a fixed percentage each time. For example, with the data above, 20 roach were used. One roach therefore represents 5% of the sample. This figure is then halved to give 2.5% and is removed from each of the cumulative percentage mortalities. The original 5% mortality after 5960 min becomes 2.5%, the 10% becomes 7.5% and so on up to 100%, which becomes 97.5%. This method also allows an extra point to be plotted on the probit paper, though most attention should still be directed to the values in the range 40%–80%. The probit in Figure 14.3 shows the result of the

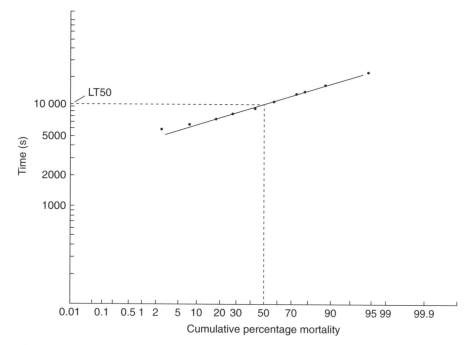

Figure 14.3 Probit of roach toxicity data showing the estimation of the LT50 corrected for extreme responses.

transformation. The approximate straight line relationship is retained but the LT50 is increased to 10 650 min.

Confidence intervals can also be estimated for the LT50 using the probit paper. Values of LT16 and LT84 are first obtained from the plot. These are then used to generate a value of s

$$s = (LT50/LT16 + LT84/LT50)/2$$

The s value is then substituted in the formula for f

$$f = antilog[(1.96 \log s)/\sqrt{n}]$$

where n is the number of animals in the tank. The 95% confidence intervals are obtained as

$$LT50 \times f \quad \text{for the upper interval}$$

$$LT50 \div f \quad \text{for the lower interval}$$

Other confidence intervals can be obtained by substituting different values of z (**1.96**) in the equation. Quite often in such studies, not all of the organisms die in the test tank over the period of study. When this occurs, the value of n in the confidence limit calculation is not the total number of animals in the tank, nor the number which have succumbed, but a number in between. Litchfield (1949) provides a method for obtaining n in these cases.

BOX 14.1: 95% CONFIDENCE LIMITS TO THE LT50

For the roach example, the LT16 and LT84 values from the second probit analysis are 7250 and 15 500, respectively, so that

$$s = (10\ 650/7250 + 15\ 500/10\ 650)/2$$

$$= (1.469 + 1.455)/2$$

$$= 1.462$$

$$f = \text{antilog}[(1.96 \log 1.462)/\sqrt{(20)}]$$

$$= \text{antilog}(0.3233/4.472)$$

$$= \text{antilog}(0.0723)$$

$$= 1.181$$

giving 95% confidence intervals of

$$LT50_u = 10\ 650 \times 1.181 = 12\ 580$$

$$LT50_L = 10\ 650/1.181 = 9020$$

Note that the intervals are not symmetric about the LT50.

Table 14.2 Acute cadmium toxicity data for blue gill fry (24 h, $n = 20$ for each tank)

Cd (mM/l)	Fry alive after 24 h (%)
0.0	100
0.052	95
0.060	85
0.070	65
0.081	40
0.10	20
0.14	5
0.20	0

14.3 The LC50

Experiments can also be performed to obtain an estimate of the median lethal dose (LC50). Here the set-up differs from that used for the LT50, though probit analysis may again be used to obtain the estimates. In Table 14.2 are some data relating the survival of blue gill fish fry to cadmium ions. Here the experiment is designed so that separate samples of fish are exposed to a range of cadmium concentrations over a fixed time. The period of time will depend upon both the nature of the toxicant and that of the test organism,

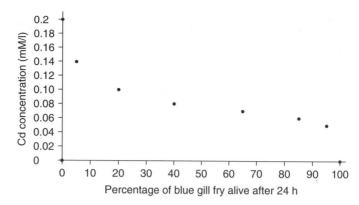

Figure 14.4 Percentage survival of blue gill fry plotted against cadmium concentration.

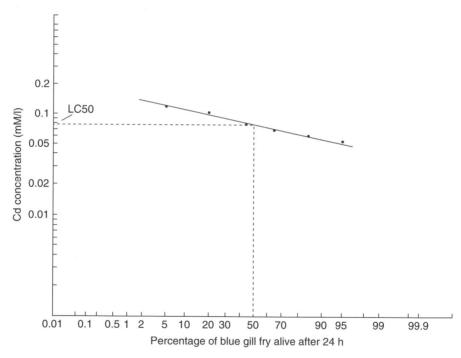

Figure 14.5 Probit of blue gill data showing the estimation of the LC50.

but is usually less than one week. In this case the fish are exposed for 24 h and in each tank the percentage survival is recorded.

The percentage survival is plotted against the cadmium concentration in Figure 14.4. Again note the sigmoid curve, which can be linearised using a combination of a logarithmic transformation and probability paper.

After ensuring that the probit paper is correctly oriented as in the LT50 example above, the LC50 is obtained by plotting the percentage survival along the probability axis and the cadmium concentration along the logarithmic axis, as shown in Figure 14.5. The 50% survival value is again projected to the best fit line and the LC50 is read off as 0.085 mM. Again, not all the data in the table can be plotted. The 'control' cadmium concentration of 0 mM/l cannot be located on a logarithmic scale. Neither can the 0% survival be plotted on the probability scale (the point lies at infinity).

It should be apparent that the methods applied to obtain the LT50 and LC50 provide another example of modelling. In this case the cumulative normal distribution curve is used to simulate the relationship between the toxicant and the organism, similarly to the way in which the Gompertz curve in Chapter 13 was used to estimate the growth rates of mammals.

There are several other methods which can be used to analyse toxicity data. One of the most popular employs 'logits' rather than probits. Logits are obtained from the logistic equation, an exponential form which also produces an S-shaped curve like that obtained in many, but not all, toxicity experiments. The logistic equation has more convenient statistical properties than those associated with the probit method, but no simple graphical method is available. However, a number of statistical packages allow linearisation using the logistic equation (e.g. Minitab and Statistical Analysis Systems, SAS) and produce estimations of the LC50 and confidence intervals. The reader is referred to Piegorsch and Bailer (1997) for examples of its use. For further information on probit analysis consult Bliss (1937), Finney (1971) and Hewlett and Plackett (1979).

Key notes

- Toxicity testing frequently makes use of the cumulative normal distribution (probit) to estimate the statistics LT50 and LC50.

- The LT50, or median survival time, is a fixed temporal measure of an organism's response to a toxicant.

- The LC50, or median lethal dose, is a fixed concentration of toxicant which leads to death of half of the organisms exposed during a specified time.

CHAPTER FIFTEEN

An introduction to multivariate analysis

A statistical definition of a variable is 'a quantity which may take any one of a specified set of values'. Many examples of variables have been given in the preceding chapters, including continuous variables, such as temperature, and discontinuous variables, such as traffic density along a road. Non-measurable variables, such as colour, which can be expressed on a nominal scale, also exist. Statisticians make a distinction between variables and **variates**. A variate is a quantity which may take any of the values of a specified set with a specific frequency or probability.

Until now, only univariate and bivariate data sets have been described and analysed. Examples of analysis of the former include confidence intervals for the mean, and of the latter the techniques of correlation and regression. It should already be apparent that univariate methods with large sample numbers (e.g. ANOVA) and bivariate methods require more complex calculations than univariate methods with small sample numbers (e.g. the t test). It will come as no surprise to learn that multivariate methods, where variables number three or more, are computationally much more demanding. Situations often arise where a large number of variables have to be considered. Indeed, even a small volume of the upper ocean contains a large number of quantities which vary in time and space, such as the different species of planktonic plants and animals comprising the flora and fauna, together with numerous environmental variables, such as temperature, salinity, pH and so on. It defies any generalisation and must be considered as a dynamic multivariate system. To understand the natural environment it has been customary to effect controls to reduce the number of variables by treating most of them as constants – hence the use of laboratory simulations using constant environment rooms. This enables a researcher to focus upon a small number of variables, typically one or two, which can be analysed using conventional statistical methods. Multivariate analysis offers the possibility of analysing natural and

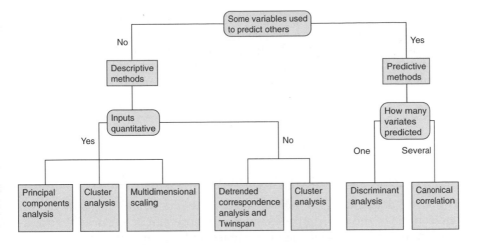

Figure 15.1 Multivariate methods (modified after Jeffers, 1978).

experimental systems where typically one may identify 20 or more measurable variables.

A wide range of multivariate techniques is available catering for many types of applications. In common with univariate methods, it is important to become familiar with the methods before using them in a research application. Unfortunately, it is not possible to provide more than a short summary of the most widely used methods here, but their increasing use in environmental science justifies their inclusion even in introductory texts. Multivariate methods have been applied to a wide range of environmental problems, including the effects of pollution on ecosystems, the description and classification of communities, land use (including remote sensing), biological monitoring, environmental impact, the distribution of organisms and pollutants, climate change and the analysis of questionnaires. To master multivariate methods requires a working knowledge of matrix algebra to first year university level. This will not be available to most environmental science students, but some basic knowledge of the techniques is not difficult to acquire and will be of great advantage if multivariate methods are to be used. For an introduction to these techniques, the student is referred to Manly (1994). Further matrix algebra is provided by Namboodiri (1984) and Nash (1990). A good introduction can be gained in a few weeks of study, beginning with an A level in mathematics or equivalent qualification.

It is important to note that many multivariate methods are not inferential. The aim is often not to test hypotheses but to generate them from the data. With large numbers of variables a first step is usually to provide a summary of the data by clarifying the relationships between the variates and/or samples from which they were taken.

A diagram of the main multivariate methods and their relationships is shown in Figure 15.1, a flowchart adapted from Jeffers (1978). Two classes of multivariate analysis are indicated. The first includes descriptive methods, which clarify the underlying data structure. The second includes predictive techniques, which provide a means of classifying new multivariate data sets into a number of categories. Both methods produce tabulated and graphical outputs. Graphical results include dendrograms (tree diagrams) and scattergraphs, allowing the relationships between the many variables and/or samples to be visualised.

15.1 Descriptive methods

15.1.1 Principal components analysis

The method known as principal components analysis is the best known and earliest of the descriptive techniques; whose aim is to clarify the variation within a series of multivariate samples and display samples in a scattergraph so that similar samples appear close together and dissimilar samples far apart. The pattern of the scattered points can also demonstrate environmental (or other) trends, which may lead to explanations for the variation between the samples. In common with the other descriptive techniques, principal components analysis is used on fairly large samples (>10) and a large number of variables (>10). The variables should be continuous, but discontinuous and ordinal measurements are often used, though the interpretation of results may then be ambiguous. With principal components analysis it is possible to observe correlations between all the variates and to clarify relationships between the samples. This is achieved through the calculation of a set of principal components, which effectively reduces the number of variables to a small number of new variables, which are uncorrelated. This is best understood with reference to Figure 15.2(a), where a scattergraph shows a correlation between two variates. The points have an elliptical distribution and the direction of greatest variation is shown by the line PC1. A second line, PC2, has been drawn at right angles to PC1. If these new axes are then used to plot the scatter of points, by rotating them as shown in Figure

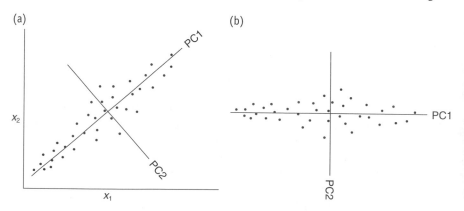

Figure 15.2 Positioning of principal components PC1 and PC2 for a simple two-variable example.

15.2(b), it is clear that there is no correlation between the new 'variables' labelled PC1 and PC2.

Principal components analysis draws new axes using a computer program to maximise the variance in a particular direction. With two variables the calculations are straightforward, but with many variables they are much more difficult. Measurements of ten variables would need ten dimensions to plot the samples in a hyperspace, which cannot be visualised. Ten principal components can be found, but usually only the first few components (PC1, PC2, PC3) are significant as they account for most of the variance. This is the advantage of the technique – a small number of principal components are obtained and the samples are plotted along them, allowing the relationships between the samples to be seen. This procedure of reducing the 'dimensionality' of the data and plotting it along a small number of axes is called **ordination**.

In the raw data it is usual for a considerable number of the variables to be significantly correlated. For example, in soil samples, a correlation between the amount of exchangeable potassium and the clay content is to be expected because clays are strong adsorbers of potassium ions. In fact, if none of the variable pairs is significantly correlated, a principal components analysis is not justified. Computer programs allow a researcher to view the correlations before an analyis is undertaken. They could be calculated by hand, but for a 10 samples × 10 variables case, 45 correlation coefficients would need to be calculated!

Jeffers (1978) provides an instructive example of principal components analysis as part of an environmental impact assessment. Here 274 samples of marine sediment from a bay were analysed to assess the impact of a proposed barrage. Each sediment was analysed for percentage particle size (four categories) and nutrient content (four categories), giving a total of eight environmental variables. Significant correlations were obtained between several variable pairs, but even with eight variables it was not possible to detect any patterns in the samples, so principal components analysis was used for clarification. A primary matrix was first prepared. A matrix is similar to a spreadsheet and in this case it consisted of 274 rows, one row for each sample, and eight columns for the variables. Each cell (matrix element) contained one measurement, and a computer package was then used to perform the analysis.

The output listed the ten principal components with the amount of variance extracted by each of them. In this example it was found that the first four principal components extracted 88% of the total variance, so the eight original variables had effectively been reduced to four. Each site can be plotted against these components to view their relationships with each other. In this case the first component (PC1) was found to represent sediment fertility and the second (PC2), a measure of the biological calcium carbonate content. Contour maps can then be plotted using the values of the principal components to indicate regions of high fertility and carbonate content within the bay. The analysis was taken further by measuring the abundance of the major invertebrates inhabiting the sediment samples.

Table 15.1 Example of a small correlation matrix

		Variable			
		1	2	3	4
Variable	1	1.0			
	2	−0.34	1.0		
	3	0.78*	0.16	1.0	
	4	0.32	−0.04	0.11	1.0

Macoma balthica

These consisted mainly of molluscs and worms, which were placed into 22 categories (22 biological variables) for a second principal components analysis which was carried out independently of the environmental variables. Again, correlations were observed between certain species, and the principal components analysis showed that the first five components accounted for 86% of the variance within the sample. The first component was found to be an indicator of the abundance of three common invertebrate species. Finally, some important information was obtained by correlating values obtained from the two principal components analyses for each site. For example, it was found that the numbers of *Macoma balthica*, *Hydrobia ulvae* and *Nereis diversicolor* were positively correlated with sediment fertility. If a barrage were to be constructed across the bay, changes in the sediment rain would occur due to the altered hydrography; the multivariate analyses provide a valuable summary of the sediment characteristics and will help predict changes that will occur to the animal populations.

Principal components analysis is not normally suitable for data sets where the number of variables is less than about six, unless the number of samples is very large. In these cases, an alternative is to obtain the correlation coefficient between every pair of variables. The results are best displayed as a matrix, as illustrated in Table 15.1 for four variables.

Matrix elements (cells) along the diagonal all contain coefficients of value 1, since the variables are perfectly correlated with themselves. The only correlations of interest occur in the elements below the diagonal. Of the six possible combinations, only one correlation (variables 1 and 3) is significant. With larger numbers of variables caution must be exercised. As was explained in Chapter 10, analyses performed repeatedly in this fashion are bound to turn up 'significant' results by chance, leading to the false conclusion that a significant correlation occurs. An interesting example of

Table 15.2 Diatom relative frequencies at twelve river sites

Site	s1	s2	s3	s4	s5	s6
1	85	64	76	51	18	0
2	93	78	31	56	22	12
3	46	81	88	42	31	12
4	22	77	42	13	47	35
5	41	49	32	18	41	44
6	28	55	61	56	28	86
7	13	61	13	52	16	71
8	22	14	27	32	45	68
9	18	12	15	17	31	69
10	4	10	1	44	58	63
11	21	28	1	31	46	62
12	10	5	0	30	59	85

spurious correlations obtained with multivariate analysis is given by Rextad *et al.* (1988).

Factor analysis is a related method available in several statistical packages. The method has received considerable criticism on the basis of its subjectivity – the user decides on the form of the analysis to be undertaken, so the results may depend as much on the user as on the data. Despite this, the method is still fairly widely used in research.

Principal components analysis using Minitab

A hypothetical example is given consisting of a table of diatom relative frequencies collected from 12 sites along a river. Data for six species of diatom, s1–s6, are given in the columns of Table 15.2. Several freshwater diatoms have been shown to be sensitive to organic pollution. Principal components analysis is used to ordinate the data in the table along the first two principal component axes.

The measurements are entered into columns 1–6 of a new worksheet. On the menu bar:

Stat > Multivariate > Principal Components

In the dialog box, under 'Variables' type C1–C6. The box labelled 'Number of components to compute' can be left blank, since there will only be a maximum of six. If the number of variables were larger, only the first five to six components need to be specified. Use 'correlation matrix' (the default) for this analysis.

Now click 'Storage coefficients' and under 'Scores' type C10–C15. You must assign sufficient columns by ensuring that the column number is at least as great as the number of variables in the data (in this case six). Click 'OK' twice. The output appears below.

Eigenanalysis of the Correlation Matrix

Eigenvalue	3.7338	1.0015	0.5233	0.3848	0.2630	0.0937
Proportion	0.622	0.167	0.087	0.064	0.044	0.016
Cumulative	0.622	0.789	0.876	0.941	0.984	1.000
Variable	PC1	PC2	PC3	PC4	PC5	PC6
C1	0.453	−0.087	0.572	−0.097	0.250	−0.623
C2	0.448	−0.137	−0.381	−0.118	−0.736	−0.282
C3	0.418	−0.191	−0.518	0.532	0.486	−0.034
C4	0.272	0.775	0.211	0.469	−0.211	0.126
C5	−0.404	−0.383	0.242	0.686	−0.314	−0.250
C6	−0.427	0.435	−0.395	−0.055	0.129	−0.673

	C10	C11	C12	C13	C14	C15
1	3.23225	−0.07621	0.56771	0.17395	0.663927	0.223699
2	2.74354	0.44448	1.30619	−0.40412	−0.459721	−0.335822
3	2.35542	−0.72678	−0.71533	0.76790	−0.014778	0.330476
4	−0.01732	−1.86359	−0.77491	−0.13058	−0.727431	−0.084228
5	−0.16562	−1.18642	−0.00367	−0.40839	0.083903	−0.279223
6	0.54602	1.48970	−1.11315	0.59234	0.241643	−0.513004
7	0.20003	1.70936	−0.70386	−0.89835	−0.584143	0.325950
8	−1.28931	0.01490	0.09529	0.27214	0.551741	−0.020230
9	−1.44425	−0.25320	−0.19615	−1.03068	0.877419	0.194906
10	−2.07978	0.44043	0.68558	0.86059	−0.393739	0.431452
11	−1.41487	−0.04499	0.43045	−0.22110	−0.272326	0.007492
12	−2.66611	0.05233	0.42184	0.42631	0.033506	−0.281468

The eigenanalysis provides for each principal component its eigenvalue, its proportion and its cumulative proportion. The eigenvalue is a measure of the amount of variance extracted by a given component – the larger the value, the more variance extracted. In this case most of the variance is accounted for by PC1 since its eigenvalue is nearly four times as high as the eigenvalue for PC2. PC2 also accounts for considerably more variance than the remaining principal components.

Listed below the eigenvalues are the eigenvectors (coefficients) for each PC. The individual values of these quantities allow us to interpret which variables are exerting most influence on the principal components. For example, in the column labelled PC1, coefficients C1–C3 are all positive and C5 and C6 are negative but of similar magnitude, indicating a contrast in these species. A small size of the coefficient for C4 indicates that its weighting is low and s4 has less ecological influence than the other species. For PC2, the coefficient C4 is the highest in magnitude and has the greatest weighting so that this component is most strongly influenced by species s4.

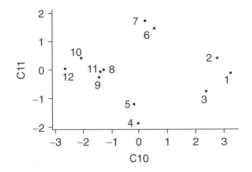

Figure 15.3 Principal components ordination.

The scores for each sample are revealed in columns C10–C15 of the current worksheet.

Ordinations can be displayed for selected principal component pairs. Since PC1 and PC2 extracted most of the variance, the scores for these components (listed in C10 and C11) are most likely to produce a useful pattern. To reveal the ordination, go to the menu bar:

Graph > Plot

In the dialog box place C11 in the top left-hand box ('Graph 1, X') and C10 in 'Graph 1, Y'. Then click 'OK'. The output is shown in Figure 15.3.

The spread of points along axis PC1 is caused by a response to a hypothetical pollution gradient from left to right. Sites 1–3 are dominated by the 'clean water' species s1–s3, whereas sites 10–12 contain an abundance of pollution-tolerant species s5 and s6; s4 appears to be unaffected by the water quality.

Diatom species s1–s3 show a pronounced decline in abundance from sample 1 to 12 (see Table 15.2), while s5 and s6 show an increase as a response to the pollution gradient.

15.1.2 Correspondence analysis

Detrended correspondence analysis is currently one of the most popular multivariate methods used by ecologists (Hill and Gauch, 1980). The analysis produces similar results to principal components analysis but using different computational methods. The multivariate data may be of the 'presence/absence' type rather than discrete measurements, which makes the method particularly suitable for ecological work. However, with a simple modification, quantitative data may be included to provide a useful general method of analysis. Detrended correspondence analysis may also provide a more interpretable ordination than principal components analysis and correspondence analysis as it is not subject to the 'arch effect' – a distortion common to several ordination methods (Gauch, 1982).

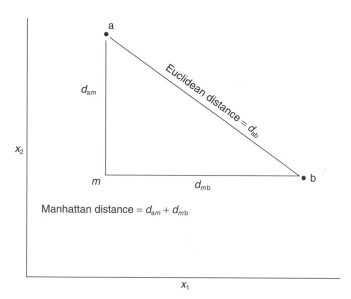

Figure 15.4 The Euclidean distance.

15.1.3 Multidimensional scaling

Multidimensional scaling originated in psychometric research (Kruskal and Wish, 1978), but has been successfully applied to a wide range of environmental problems. The principle is based upon finding the best relationship between the inter-point distances on an ordination with some mathematical measure of similarity between each of the samples. Each sample, containing a number of measured variables, is compared with every other sample by this similarity or 'distance' measure. If we take a simple case with just two measurements on each sample, the samples can be plotted as a simple scattergraph with the two variables represented by the two axes of the graph (Figure 15.4).

According to Pythagoras's theorem, the measure of similarity known as the Euclidean distance (d_{ab}) is given by

$$d_{ab} = \sqrt{[(X_{1b} - X_{1a})^2 + (X_{2b} - X_{2a})^2]}$$

With multivariate data the Euclidean distance between two sample pairs can be obtained using an extension to this equation, but it cannot be graphed as it occurs in multidimensional space

$$d_{ab} = \sqrt{\left[\sum_{i=1}^{m}(X_{ia} - X_{ib})^2\right]}$$

where m is the number of variables.

A computer package allows the distances between sample pairs to be calculated without effort. With these calculations it is possible to produce a matrix of Euclidean distances between each sample pair. The matrix of

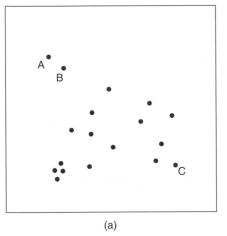
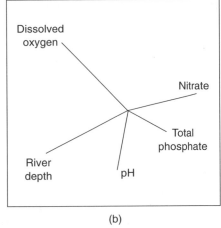

Figure 15.5 (a) Non-metric multidimensional scaling ordination of benthic invertebrate samples from a river; (b) environmental correlation diagram.

distances is then compared with a trial ordination of the samples as points on a scattergraph in two or three dimensions, until a best fit is obtained. This is more difficult and there are several computer methods to solve the problem. One of the most useful variants of multidimensional scaling is non-metric multidimensional scaling, where the data assumptions are relaxed and the similarity distances are ranked rather than scaled to obtain the ordination. A hypothetical example of an a non-metric multidimensional scaling ordination is shown in Figure 15.5 for fly larva communities in a polluted river. The ordination in Figure 15.5(a) shows the spatial relationships between 18 benthic samples. Samples A and B are close together indicating a high degree of similarity, while A and C are very dissimilar. In Figure 15.5(b), the direction of correlation is shown for environmental variables measured at each site. This figure indicates relationships between the larval communities, which can be used to determine which (if any) of the variables is responsible for the differences between the sites. Multidimensional scaling, and its hybrid variants using a range of similarity measures, has been found superior to both principal components analysis and detrended correspondence analysis in some applications and its popularity is increasing.

15.1.4 Cluster analysis

Cluster analysis encompasses a wide range of multivariate methods, which result in the grouping or clustering of similar samples and the separation of dissimilar samples. It is allied to ordination but the output normally consists of a **dendrogram** which shows the relationships between all the samples. Analyses producing dendrograms are termed *hierarchical* and they are currently the most popular clustering techniques. All cluster analyses require the calculation of a distance or correlation matrix, such as those described

Table 15.3 Invertebrate data for cluster analysis

Sample	A	B	C	D	E	F	G	H	I
1	45	0	30	32	3	0	22	0	2
2	2	0	45	21	15	0	14	0	0
3	1	0	40	15	35	0	4	0	0
4	0	0	5	26	0	15	0	9	2
5	2	0	2	28	0	3	0	3	3
6	3	23	2	8	0	1	0	2	32
7	23	3	4	55	0	0	0	0	4
8	4	0	20	8	30	0	5	0	1

Key: *A Asellus aquaticus* (crustacea); *B Chironomus* (fly larva); *C Gammarus pulex* (crustacea); *D Glossiphonia complanata* (leech); *E Hydropsyche* (fly larva); *F Lymnaea stagnalis* (mollusc); *G Physa fontinalis* (mollusc); *H Rhyacophila* (fly larva); *I Tubifex tubifex* (worm).

Chironomus

above. For binary (presence/absence) data, an association coefficient can be used (e.g. the Jaccard coefficient). There are a considerable number of distance measures and choosing the most appropriate may be difficult. When the number of distance/correlation measures is multiplied by the range of 'linkage' methods, the number of combinations becomes large, and in general it is not possible to make a 'best choice' prior to data analysis. Some guidance is available, but the ultimate choice is still largely a subjective matter.

A useful and general method of cluster analysis for quantitative data is *average linkage*. Here, if the distance measure between samples A and B is 1.8 and the distance between A and C is 2.9 then the distance from A to BC is calculated as $(1.8 + 2.9)/2 = 2.35$. If the clusters are larger than two the calculations become more complex and must be entrusted to an algorithm.

Cluster analysis with Minitab
The data presented in Table 15.3 were obtained from 'kick' samples in an English river polluted by sewage effluent. A series of sites was sampled above and below a sewage outfall and the major invertebrates were quantified on a relative scale. A cluster analysis will be used to classify the sites and determine whether sewage pollution influences the number and species of invertebrate present. In Table 15.3 the invertebrate species/genera are placed in columns and the samples in rows.

A glance at the table shows that some species are abundant and widespread (e.g. the leech *Glossiphonia complanata*) while others are less common and local (e.g. the bloodworm *Chironomus*).

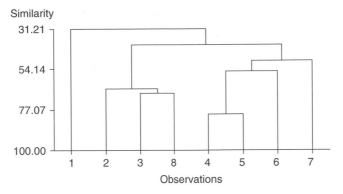

Figure 15.6 Dendrogram for an average linkage cluster analysis (Minitab).

The table is entered into a new Minitab worksheet with eight columns, C1–C8, containing the species relative abundance. On the Minitab menu bar:

Stat > Multivariate > Cluster Observations

In the dialog box enter C1–C8 in the 'Variables' or 'Distance matrix' box. Highlight 'single' in the 'Linkage method' box and 'Euclidean' in the 'Distance measure' box. The variables do not need standardising as they are all relative abundancies. The number of divisions in this case will not be specified. Now click 'show dendrogram' followed by 'OK'.

The dendrogram resulting from the analysis is shown in Figure 15.6. The sites have been linked together in a manner which allows us to see their interrelationships. Site pairs (3, 8) and (4, 5) are the most similar and there is a clear separation of sites into three groups, with site 1 in the first group, sites 2, 3 and 8 in the second group, and sites 4, 5, 6 and 7 in the third group. The analysis is revealing because the sites in the last group were all taken below the sewage outfall where there was additional evidence of pollution shown by the presence of some sewage fungus and a high biological oxygen demand (BOD). Site 1 was taken much further down the river in the 'recovery zone', where a number of pollution-sensitive species began to reappear. Sites 2, 3 and 8 were above the sewage outfall, where there was no evidence of organic pollution. The close association between sites 4 and 5 results from their proximity – they were only 40 m apart. If you look at Table 15.3 you will see that these two sites have most species in common and at similar relative abundance. An almost identical dendrogram is obtained using the Manhattan distance in the 'Distance measure' box (see Figure 15.4). The analysis has clearly distinguished between the sites and provided a classification which is related to water pollution.

Another popular clustering technique is TWINSPAN (Hill, 1979). This method has many useful applications and is performed on ordinations obtained by detrended correspondence analysis. The method classifies both samples and variables and the analysis can be stopped after a chosen number of divisions. It has been used successfully to classify unpolluted river faunas in the UK (Wright *et al.*, 1984; Clarke *et al.*, 1996) and has been developed to enable new sites to be compared with an unpolluted 'control' set. Since the UK's rivers differ widely in their physical, chemical and biological characteristics, inputting data covering these characteristics allows a researcher to compare the observed fauna of a new river sample to the 'expected' pristine fauna using the computer package RIVPACS. If there is a serious mismatch the river may have been polluted.

15.1.5 Other descriptive measures

Further methods employing ordination include principal coordinates analysis, which is a more generalised form of principal components analysis, and detrended canonical correspondence analysis (ter Braak and Verdonschot, 1995). Several other techniques have been used from time to time with varied success. It should be noted that ordinations can be sensitive to both the nature of the input data (standardised or raw) and the distance measure employed. Extensive studies of a wide range of descriptive methods using artificial data have led to the conclusion that, for both ordination and cluster analysis, the application of several methods to the same data set is the most sensible policy (Kenkel and Orloci, 1986). Table 15.4 provides some references to ordination techniques with some applications relevant to environmental science.

Table 15.4 Literature relating to the main methods of ordination

---

Canonical correspondence analysis
ter Braak and Verdonschot (1995)

Canonical correlation analysis
Giffins (1985)
Christie and Smol (1993)
ter Braak (1994)

Detrended correspondence analysis
Hill and Gauch (1980)
Whittacker (1987)

Principal components analysis
Minchin (1987)
Manly (1994)

Multidimensional scaling
Faith *et al.* (1987)
Minchin (1987)
Newell *et al.* (1990)

15.2 Predictive methods

Predictive methods differ from the preceding methods in one important aspect – one or more sets of variables can be predicted or distinguished from other sets. Many of the descriptive methods described above are becoming modified to include a predictive option, however, so the distinction is not as clear as it once was. The first of these methods, *discriminant analysis*, aims to find the equation of a line separating two subgroups of samples and allocate future samples to one of the two groups. The method works best where there is a fairly sharp boundary between the groups, though this is rarely observed in nature. As an example, it may be of interest to classify lake waters on the basis of their chemical characters in order to divide them into two groups – one vulnerable to acidification and the other well buffered against acidification. 'Calibration' samples would need to be collected to include as large a range of lakes as possible. After the analysis it should then be possible to assign other lakes to one or other of the categories. The analysis would provide a useful classification tool which can be used in lake management. No prior knowledge is needed for the choice of variables, but methods are available to test which variables contribute most to the discrimination so that the number of variables to be included in the analysis can often be reduced at a later stage, saving time and money. The method is computationally demanding and requires multivariate normal distributions. It can also be extended to classify samples into several groups rather than just two.

The second method, known as *canonical correlation analysis*, enables one set of variables to be correlated with a second set, where all of the variables have been measured on a series of samples. It is a potentially powerful method for identifying which environmental variables are responsible for biological variation in the field. The method has some relationship to principal components analysis, but maximises the correlation between two groups of variables. Canonical correlation could be applied to two of the examples mentioned above. In the environmental impact study, the marine sediment characteristics could be compared with the biota of these sediments. The relationships can be tested statistically and it is often possible to determine which environmental variables contribute most to the biological structure. In the river pollution survey, the invertebrate samples could be correlated with characteristics of the river sediments or the water chemistry at the sites. The method is useful where sets of related variables are measured on two or more major components of an ecosystem. Ordination of the data is also achievable. The method has been criticised because frequently the underlying model assumptions are violated and the interpretation of the analyses then needs to be undertaken with care (Gauch, 1982). The environmental variables must be independent (uncorrelated) and transformations may be necessary prior to the analysis. Canonical correlation is available in several packages, e.g. the Statistics Package for the Social Sciences (SPSS). For further information see Table 15.4.

15.3 Significance testing with several variables

Much of multivariate analysis is descriptive rather than inferential, and the results of an analysis often lead to an appreciation of data structure, which can subsequently be used for the basis of a well-defined field sampling programme. There are also situations where the number of variables measured on samples are too small to use any of the above methods and yet a test is needed to subtantiate a hypothesis. For example, we may have a sample of ten soils from field A and ten from field B. If we measure just one variable on each of the soils, such as organic matter content, we might compare A and B using a Mann–Whitney or a t test. If, however, we measured several variables on each of the soils the test becomes more complex. One possibility is to take the variables one at a time and test A with B as above. This raises an objection which has already been discussed in relation to ANOVA (Chapter 10). An overall test of the variable means is more appropriate. Hotelling's T^2 test is a widely used method. The test is related to the t test and assumes normality and equal within-sample variance. A computer program is needed to perform the test as it requires matrix manipulation.

Other methods are available for more complex sets of data. We might have obtained soils from three or more fields while measuring several variables on each soil, such as pH, organic matter content, exchangeable K and exchangeable Ca. Each field may be considered a 'treatment' and in this case Bartlett's test may be used to analyse the relationships between variables and treatments. Examples of these methods are provided by Manly (1994).

Key notes

- Multivariate methods are used for the simultaneous analysis of a series of samples with many measured variables.

- Descriptive and predictive methods can be distinguished.

- Currently popular descriptive methods are principal components analysis, detrended correspondence analysis, multidimensional scaling and cluster analysis.

- Principal components analysis, detrended correspondence analysis and multidimensional scaling can be used to produce ordinations – scattergraphs clarifying relationships between the samples or variates.

- Most cluster analyses produce dendrograms which are used to classify samples or variates.

- Predictive methods include discriminant analysis and canonical correlation analysis.

- Predictive methods use one set of the variables to predict the values of another set of variables. They can be useful in environmental studies where both biological and environmental data are collected from the same sites.

CHAPTER SIXTEEN

Questionnaires

An efficient way of collecting data about people's habits, awareness and environmental understanding is the questionnaire. Information collected by a questionnaire may be statistically analysed or used as a purely descriptive tool in social research. Data from questionnaires are frequently reported and raise environmental awareness, and may even influence political decisions. The construction of a good questionnaire requires skills in both language and statistical design. Considerable thought is required in the planning stage, and care taken in question design is invariably repaid later. One of the commonest mistakes made in undergraduate questionnaires is to concentrate on a small number of issues while ignoring the methods of analysis. Consequently, some interesting topics can be covered but the results may be impossible to analyse so that clear, unbiased statements can be made about the sample. A good questionnaire will address specific points, one at a time. It will be comprehensive without being too complex or lengthy. It is sensible to include a pilot study to iron out inconsistencies prior to distribution.

Much has been written about questionnaire design. Here, attention is drawn to the self-administered closed-question form. Such questionnaires may be administered verbally, by telephone or by mail to a sample of the population. They usually have a matrix-style format, though this is not appropriate for all types of questions.

Social scientists recognise several types of question. There are *factual* and *opinion* questions. Examples of factual questions are an individual's age, gender, height and profession. Attitudes are revealed in what the respondent thinks is desirable or acceptable. They may be distinguished from *beliefs*, which cover those things people believe to be true or false, irrespective of whether in fact they really are. *Behaviour* covers aspects of what people do in a given situation. These forms are illustrated in the following series of questions relating to domestic waste recycling.

1. Do you consider it acceptable that infirm people should segregate their kitchen waste for separate collection?   [yes] [no] *opinion*
2. Do you take your empty bottles to a bottle-bank for recycling?   [always] [occasionally] [never] *behaviour*
3. Are you aged 65 or over?   [yes] [no] *factual*
4. Does your local authority operate a bottle-recycling scheme?   [yes] [no] *belief*

Such questions may form part of a questionnaire, but if they did their order would have to be changed, because if the answer to question 4 was negative then question 2 should not arise. The order of the questions must be logical, and if possible the easiest and least personal questions should be asked first. This will put the respondent in a good frame of mind, resulting in a positive attitude, which should lead to the retrieval of a reliable sample.

Part of the questionnaire might be as follows:

1. Does your local authority operate a bottle-recycling scheme?
   [no] go to Question 5
   [yes] go to Question 2
2. (a) Do you take your empty bottles to a bottle-bank? [always] [sometimes] [never]
   (b) For how many years has the bottle-recycling scheme been operating?
       [ ] [don't know]
   (c) Were you consulted about the scheme prior to its operation?   [yes] [no] [can't remember]
       etc.

The wording of the questions is important. The following points should be borne in mind.

1. Questions should be *easy to understand*, avoiding long or little-used words, technical terms and negatives. As an example of the latter, the question 'do you think that cardboard should not be recycled?' should be replaced by 'should cardboard be recycled?' In a television survey, the question 'What proportion of your evening viewing time do you spend watching news programmes?' was a complete failure – hardly any respondents knew what 'proportion' meant and only one of the 246 respondents knew how to calculate it. It is advisable to avoid mentioning percentages for the same reason.

   Simple questions need not necessarily be short, and some researchers have found that two iterations of the same question *in the same sentence* can provide a better response.
2. Questions should *avoid ambiguity*. The question 'Do you recycle your packaging?' is too vague and might refer to one or several materials, such as plastics, paper or cardboard. Terms such as 'weekday' are often misinterpreted – some people include Saturday, while others do not.
3. *Reword leading questions.* 'Would you agree with Professor Binn, a leading expert on the economics of recycling, that *all* glass should be separated from household waste?' is a leading question. Many people will be inclined to agree with an 'authority' irrespective of their own views. 'Do

you think that all glass should be separated from household waste for recycling?' should provide a less biased response.

4. *Sensitive issues should be worded carefully.* The question 'How often have you thrown glass bottles from a car window?' could be reworded as 'How do you dispose of empty glass bottles on road journeys?' [return home] [leave in vehicle] [leave on roadside] [other]. A facile example, but sensitive issues often need to be addressed. People are easily offended, even if the questionnaire is anonymous. An alternative approach, which has proved fairly successful for face-to-face interviews, is to use the 'randomised response technique'; where the respondent is left to choose one of several questions using a random device and answers the chosen question without the interviewer knowing which question it is. The method is described by Warner (1965). For some amusing but neverthe-less instructive discussion of sensitive questions consult Barton (1958).

16.1 Open and closed questions

Compare (a) and (b) below, which attempt to obtain the same answer.

(a) Which categories of domestic waste would you be willing to recycle?

(b) Which of the following categories of domestic waste would you be will-ing to recycle? (Please circle)

newspaper	plastics
cardboard	garden waste
glass	kitchen waste
aluminium	
clothing	

Question (a) is an open question, which allows a respondent to answer with any choice of words.

Question (b) is a closed question, which prompts and guides the respon-dent into answers suggested by the researcher. In general, the closed form is the most suitable for questionnaires. It helps to organise the respondent's thoughts on the topic, some aspects of which might have been forgotten when answering in an open format. Filling in the questionnaire immediately be-comes easier, particularly for less literate respondents. Where the question-naire is given verbally, shy or inarticulate people are more likely to provide an answer to a closed question. Consequently a larger and more represent-ative sample of the population can be included. Finally, the closed form is more easily coded for analysis. Despite these advantages there is a down-side. Respondents are 'moulded' into a reply closest to, but not necessarily, the same as their own. They cannot qualify their statements with 'I agree, but it depends upon . . .'. To compromise, some surveyors recommend a mix of open and closed style questions.

Closed (fixed choice) questions can be formulated in several ways. In Likert-style questions a spectrum of replies is provided, one of which must be encircled or ticked as illustrated below.

'Disused quarries are better developed as nature reserves than filled with domestic waste.'

1   strongly agree
2   agree
3   cannot decide
4   disagree
5   strongly disagree

The marginal numbers are used for coding. An alternative range can be provided by

1   2   3   4   5   6   7   8   9   10
strongly   strongly
disagree   agree

Likert-style questions are widely used but are prone to the psychologist's 'acquiescence response set', where some respondents will work through a form encircling 'agree' throughout. To remove this problem, a series of statements is provided and the respondents are asked to encircle that which most closely reflects their views. Further difficulties may arise with the 'cannot decide' group. With opinion-type questions, these words may be chosen because the respondent truly has no opinion, while for factual questions the respondent simply cannot provide the right answer.

Another type of closed question is the 'ranking format'. Here the respondent is asked to rank a series of brief answers to a given question. The list should not be too long and the method is not suitable for questionnaires administered verbally. An example would be to rank the following priorities for a recycling scheme:

Efficient collection
Provision of containers
Council Tax reduction
Reliable collections
Installation of bottle-banks
Free telephone advice
Good information

The highest priority would be given a '1', followed by a '2' for the second highest priority and so on.

All questionnaires require an introduction to explain why the survey is being undertaken, who will be using the data and how it might benefit the respondent. It is important to ensure anonymity and make clear how the questions are to be answered. It is advisable to print the questions on one side of a page, allow plenty of space, particularly for open questions, and leave the more sensitive or difficult questions to last. It may be desirable to section the questions for a large questionnaire. Some skill will be needed to vary the way the questions are asked, while retaining simplicity to prevent the respondent becoming bored. For closed questions it is advisable to code the responses in the margin to speed up the analysis. It is also a good idea to give each questionnaire a number so that it can be referred to later.

16.2 Administration of questionnaires

Replies to questionnaires are normally received during 'face-to-face' interviews, by telephone or through the mail. There has been much discussion of the relative merits of these avenues of communication, but overall it appears that mailing is marginally superior, especially when costs are considered. The form of the questionnaire will clearly depend upon the way it is to be administered, and with telephone interviews the wording needs to be as concise as possible.

A particular concern for all researchers is the ratio between the number of questionnaires administered to the number returned and correctly filled in. This value, usually calculated as a percentage, (number correctly returned/ number administered) × 100, is called the 'response rate'. Response rates for all three avenues are often as high as 70%–80% for well-designed, well-targeted questionnaires, but they can be much lower. The author once sent 200 questionnaires to farms whose land contained large areas of undrained, low-lying land in the UK. The addresses were guessed from Ordnance Survey maps and the response rate was a pitiful 13%. One of the main problems was a lack of knowledge regarding the 'target'. Merely mailing to an address is inadvisable as there may be several inhabitants per household and the questionnaire could remain on the mantelpiece unanswered. Furthermore if there is no 'follow-up' letter, this can considerably decrease the response rate. These points emphasise the importance of a good administration strategy for the questionnaire.

A low response rate for mailing may also introduce an unacceptable level of bias, because the non-respondents will include a higher proportion of people with reading or writing difficulties. It is also more difficult to obtain representative samples from the general population using mailshots.

Postal surveys should always include a cover letter with a letterhead, the name and address of the respondent if known, and a brief explanation of the survey. The letter should give an assurance of anonymity and some information about the researcher. A genuine handwritten signature is probably worth the effort. Nowadays, envelope preparation must be given some thought. A business-advertising type of envelope will probably be binned. A personal touch is worth considering. Handwriting the addresses will be laborious but the letter will almost certainly be opened, particularly if an adhesive postage stamp is used. A stamped and addressed envelope must be included for the reply, which of course immediately doubles the cost of the exercise. Even the mailing time may influence the response rate – avoid public holidays and mid-summer when most people leave for holidays. Follow-up letters are also worth the effort. An initial introductory letter should be considered to warn the potential respondent to expect a questionnaire. If the questionnaire is not returned after seven–ten days, a brief and polite reminder should be sent. A second or third follow-up may be considered, but be careful to word it correctly. If the first follow-up gives a deadline for the completion of the survey, make sure that it is adhered to, so that the deadline has not been passed by the later follow-up letters.

Some surveys require detailed information of the respondent's behaviour over an extended length of time. These can be augmented with diaries supplied by the researcher to keep the time measurements reliable and avoid memory errors. Good reviews of questionnaire design and administration are provided by Kalton and Schuman (1982) and de Vaus (1996).

BOX 16.1: QUESTIONNAIRE ADMINISTRATION

The introductory letter

GG Recycling
Amadeus House
Bradlington
Sussex

tel. 0177 894535
e-mail xyz@abc.def
3rd October 1999

Janet Lalou
The Pines
Love Lane
Sleeford, Sussex

The Green Grey Recycling Group is a charity which manages recycling schemes in several counties and has been established for twenty years.

Your household has been selected to provide an opinion on recycling and inform us of recycling schemes operating in your area. Your address was selected randomly from the county list and all questionnaires will be treated anonymously.

The results of the survey will be used to assist the town council in its recycling policy. They will be presented to the town council and discussed at an open meeting next November. Full details will be provided in the local newspaper and posted outside the Town Hall. The results will also be available for you to consult at the local Reference Library.

The form, which should take no longer than five minutes to complete, will be posted to your address in seven days' time.

If you have any questions concerning the survey, please write or call. The freephone number is 080 432 6737.

Thank you in advance.

Yours sincerely
Peter Swann *Project Manager*

Letter to accompany the questionnaire

GG Recycling
Amadeus House
Bradlington
Sussex

tel. 0177 894535

e-mail xyz@abc.def
11th October 1999

Dear Ms Lalou

Last week we sent you a letter asking whether you would kindly agree to fill in a questionnaire on household waste recycling. Could you please fill in the form below. Most questions require you to draw a circle around a preferred option.

We would prefer the questionnaire to be answered by the person responsible for paying the council tax. Please reply to the address shown in the letterhead above.

Please answer all of the questions. The questionnaire is confidential and does not reveal your identity. The code number at top left is there for us to keep track of the replies.

Questionnaire (only the first question is shown)

No. 127

1. Is there a household recycling scheme operating in your area at the moment?
   [yes]   [no]   [don't know]
   etc.

Follow-up letter

19th October 1999

Dear Ms Lalou

Last week a questionnaire about recycling schemes in your area was sent to your address.

Since we have not yet received your questionnaire could you please return it in the prepaid envelope as soon as possible. This would be greatly appreciated and will allow us to forward to the Council a representative sample of the towns-people's views on this important matter.

If you have recently mailed your questionnaire please accept our apologies for troubling you again. If your form has been spoilt or mislaid, please use our freephone number (**080 432 6737**) and we will send you another questionnaire immediately.

Yours sincerely,
Peter Swann *Project Manager*

Table 16.1 Statistical methods for questionnaires

Univariate	Bivariate	Multivariate
Frequency distributions	Frequency analysis (χ^2 and proportional reduction of error)	Conditional tables Partial correlation
Descriptive statistics	Chapter 6	Multiple and partial regression
χ^2	Correlation	
Chapters 3 and 6	Chapter 11	(see also Chapter 15)
	Regression Chapter 12	

16.3 Analysis of questionnaires

Closed questions can often be coded into categories at ordinal level. For example, the following categories can be ranked:

Age 11–20 21–30 31–40 41–50 older than 51

Other answers may be treated as interval data. Age may be requested in years, though response rate and bias may reduce the amount of data retrieved. Some results can only be coded at the nominal level, e.g. gender, 'yes/no/don't know', married/single. Thus the data available from questionnaires may well contain all levels of measurement, which influences the type of statistical treatment that can be used. Some methods to analyse these data are shown in Table 16.1.

A frequent problem arises with the analysis of data sets containing two or more levels of measurements. The easiest way of dealing with this is to reduce all the data to the lowest common level of measurement. This may involve, for example, reducing an interval level to an ordinal level or an ordinal level to a nominal level. In some cases, the level of measurement of a variable can actually be ignored. This is the case with 'dichotomous variables' where only two categories are possible, as with gender. Here gender can be treated as the same level of measurement as the other variable, which may be at interval level. Many of the problems associated with levels of measurement should be sorted before the questionnaire is finally implemented. A pilot study involving friends or colleagues will often help to reveal these problems. Reducing a level of measurement invariably results in loss of information and limits the statistical techniques available.

- Questionnaires should be carefully planned using simple unambiguous language.

- Questions can be classified as factual, opinion, belief or behaviour.

- Question order must be logical, with the least sensitive questions asked first.

- Open-style questions allow the respondents to express their own views, while closed questions provide the respondent with a few carefully chosen categories.

- Closed questions can be ranked to provide an ordinal scale of measurements.

- The response rate of a questionnaire is improved by sending one or two follow-up letters.

- A wide range of statistical techniques are available to analyse well-planned questionnaires.

CHAPTER SEVENTEEN

Sampling and experimental design revisited

In Chapter 2 some aspects of experimental design were introduced with reference to hypothesis testing. After having considered some statistical methods used in the analysis of experimental results it may be useful to look again at some more aspects of this subject. In Chapter 2 a pilot experiment investigating lead uptake by the freshwater crustacean *Asellus* was considered and some of the problems encountered in experimental work were discussed. The *Asellus* experiment was a simple example where, if it was conducted correctly, the results could have been analysed using a two-sample procedure, such as the Mann–Whitney or *t* test, using an appropriate null hypothesis.

Given more time and a larger sample of animals, the experiment could have been enlarged to provide more information on the relationship between lead uptake and the presence of calcium. You might argue that such an experiment is unnecessary if the results of the pilot experiment gave a negative result, but the procedure may have been flawed. Two treatments were investigated: a 'control' where the water sample contained lead alone and a second solution containing lead + calcium. Imagine a student preparing the solutions. The test sample must contain calcium, so again a soluble calcium salt is required, but how much? The amount added needs consideration, since too much calcium could result in osmotic stress leading to altered behaviour patterns and perhaps death of *Asellus*. The outcome of the experiment may then mislead the researcher. The best procedure would be to discover the levels of calcium encountered in natural hard waters and use a representative concentration, say 40 ppm, using calcium chloride. The same should be done for lead, using lead chloride, and since lead salts are poisonous, published LC50 values for crustaceans should also be consulted.

Timing of the experiment also needs thought. The animals should not be fed during the study as this could affect the solution; neither should they be subjected to starvation, so that timing must be short compared with their metabolic requirements.

Table 17.1 Randomised block design for *Asellus* placed in beakers receiving treatments A (control, no Ca), B (10 ppm Ca), C (50 ppm Ca) and D (100 ppm Ca) under a light gradient

High					*Light intensity*				*Low*
					Block				
1	*2*	*3*	*4*	*5*	*6*	*7*	*8*	*9*	*10*
A	A	D	D	D	C	B	B	B	D
B	C	C	C	B	A	A	C	D	A
D	B	A	A	A	D	C	A	C	C
C	D	B	B	C	B	D	D	A	B

There are yet other considerations which become apparent when published work on adsorption is consulted. In particular, rates of adsorption are temperature dependent, so large differences in temperature must be avoided. This is best achieved using a constant environment chamber, but if this is unavailable a laboratory bench may need to be used. Benches rarely have an even temperature because of differences in illumination and ventilation. For instances, there may be a single light bank in the wall so that light intensity will decline in one particular direction, and since light absorption leads to warming, a *randomised block design* would be worth considering. In this case the grid is set out so that the beakers follow the gradient, as shown in Table 17.1.

Here each of the treatments, A, B, C and D, appears once in each column or 'block' and is assigned randomly across the gradient using pseudo-random numbers from a calculator. You will notice that in this case three levels of calcium plus a control are being used in the experiments and, with 40 beakers in all, we would need more than 40 *Asellus*. Placing just one animal per beaker would lead to high intrinsic variability if a natural sample containing all age groups were used. Placing five to ten randomly chosen individuals in each beaker and 'pooling' the animals during processing for lead analysis would be a better procedure in this case. The beakers are set out in the above pattern to allow the results to be analysed in such a way that the effect of the light gradient can be calculated in a two-way analysis of variance (Chapter 10) so that its significance can be determined. Occasionally, two gradients approximately at right angles may occur and in this case a *latin square* design may be used, though these tend to be more appropriate for large-scale field trials than laboratory experiments. It should be noted that even random collecting from a sample of animals sometimes presents problems. Underwood (1997) provides an example of netting fish from a tank. The researcher will be less likely to catch the more agile (fitter?) fish and these will then be underrepresented in the experiment. The same could occur in a sample of *Asellus*.

17.1 Determination of sample size

It is sometimes useful to obtain an estimate of sample size for a specified confidence interval. For example, we may wish to know how many *Asellus* we should need to estimate the mean lead concentration per animal at a particular level of calcium with an error (E) of ±5 µg with 95% confidence. This can be estimated using the equation below providing we have an estimate of the population standard deviation (σ).

$$n = (z_{\alpha/2}\sigma/E)^2$$

As an example, suppose that the sample standard deviation was 11.97 µg lead. If we require 95% confidence for the mean, then, since $1 - \alpha = 0.95$, $\alpha = 0.05$, and $z_{\alpha/2} = z_{0.025}$. Referring to **Table II** the value of $z_{0.025}$ is found by looking up in the body of the table the value $0.5 - 0.025 = 0.475$, giving a value for $z_{\alpha/2}$ of **1.96**. The sample standard deviation will be used as an estimate of the population standard deviation. This gives

$$n = (\mathbf{1.96} \times 11.97/5)^2$$
$$= (4.692)^2 = 22.02$$

Therefore, to the nearest whole number, 22 *Asellus* would be required. It is important to note that we need to be careful when estimating σ by s, and the estimate should not be based on a sample size of less than 30.

Key notes

- In an experiment, the effect of each operation on the final result must be carefully considered.

- Experimental conditions should be manipulated to cover the range of conditions in the natural environment.

- Systematic gradients occurring in large space-demanding experiments can be accommodated using a randomised block design.

References*

Barton, A. H. 1958. Asking the embarrassing question. *Public Opinion Quart.* 22: 67–8.

Bliss, C. I. 1937. The calculation of the time–mortality curve. *Ann. Appl. Biol.* 24: 815–52.

ter Braak, C. J. F. 1994. Canonical community ordination Part 1. Basic theory and linear methods. *Ecoscience* 1: 127–40.

ter Braak, C. J. F. and Verdonschot, P. F. M. 1995. Canonical correspondence analysis and related multivariate methods in aquatic ecology. *Aquat. Sci.* 57: 255–64.

Carlson, R. E. 1977. A trophic state index for lakes. *Limnol. Oceanogr.* 22: 361–9.

Christie, C. E. and Smol, J. P. 1993. Diatom assemblages as indicators of lake trophic status in southeastern Ontario. *J. Phycol.* 29: 575–86.

Clarke, R. T., Furse, M. T., Wright, J. F. and Moss, D. 1996. Derivation of a biological quality index for river sites: comparison of the observed with the expected fauna. *J. Appl. Stat.* 23: 311–22.

Everitt, B. S. 1977. *The Analysis of Contingency Tables*. Chapman and Hall, London.

Faith, D. P., Minchin, P. R. and Belbin, L. 1987. Compositional dissimilarity as a robust measure of ecological distance. *Vegetatio* 69: 57–68.

Finney, D. J. 1971. *Probit Analysis*. 3rd Edn. Cambridge University Press, Cambridge.

Gauch, H. G. 1982. *Multivariate Analysis in Community Ecology*. Cambridge University Press, Cambridge.

Giffins, R. 1985. *Canonical Analysis: A Review with Applications in Ecology*. Springer Verlag, Berlin.

Harter, H. L., Khamis, H. J. and Lamb, R. E. 1984. Modified Kolmogorov–Smirnov tests for goodness-of-fit. *Comm. Stat. Simul. Comput.* 13: 293–323.

Hewlett, P. S. and Plackett, R. L. 1979. *The Interpretation of Quantal Responses in Biology*. Edward Arnold, London.

Hill, M. O. 1979. *Twinspan – a Fortran Program for arranging multivariate data in an ordered two-way table by classification of the individuals and attributes*. Cornell University, Ithaca, NY.

Hill, M. O. and Gauch, H. G. 1980. Detrended correspondence analysis, an improved ordination technique. *Vegetatio* 42: 47–58.

Jeffers, J. N. R. 1978. *An Introduction to Systems Analysis: with Ecological Applications*. Edward Arnold, London.

Jones, J. S. 1982. Genetic differences in individual behaviour associated with shell polymorphism in the snail *Cepaea nemoralis*. *Nature* 298: 749–50.

Kalton, G. and Schuman, H. 1982. The effect of the question on survey responses: a review. *J. R. Stat. Soc. Ser. A* 1: 42–73.

Kendall, M. G. and Gibbons, J. D. 1990. *Rank Correlation Methods*. 5th Edn. Edward Arnold, London.

Kenkel, N. C. and Orloci, L. 1986. Applying metric and non-metric multi-dimensional scaling to ecological studies. *Ecology* 67: 919–28.

Kruskal, J. B. and Wish, M. 1978. *Multidimensional Scaling*. Sage Publications, Beverley Hills, CA.

Leach, C. 1979. *Introduction to Statistics*. Wiley, Chichester.

Leffler, P. E. and Nyholm, N. E. I. 1996. Nephrotoxic effects in free-living bank voles in a heavy metal polluted environment. *Ambio*. 25: 417–20.

Litchfield, J. T. 1949. A method for rapid graphical solution of time–per cent effect curves. *J. Pharm. Exp. Ther.* 97: 99–113.

Manly, B. F. J. 1994. *Multivariate Statistical Methods. A Primer*. 2nd Edn. Chapman and Hall, London.

Minchin, P. R. 1987. An evaluation of the relative robustness of techniques for ecological ordination. *Vegetatio* 69: 89–107.

Namboodiri, K. 1984. *Matrix Algebra: An Introduction*. Sage Publications, Beverley Hills, CA.

Nash, J. C. 1990. *Compact Numerical Methods for Computers*. 2nd Edn. Adam Hilger, Bristol.

Newell, P. F., Newell, R. C. and Trett, M. W. 1990. Environmental impact of an acid-iron effluent on macrobenthos and meoifaunal assemblages of the St. Lawrence River. *Sci. Total. Envmt.* 97: 771–81.

Pentecost, A. 1979. Aspect and slope preferences in a *saxicolous lichen* community. *Lichenologist.* 11: 81–3.

Piegorsch, W. W. and Bailer, A. J. 1997. *Statistics for Environmental Biology and Toxicology.* Chapman and Hall, London.

Rextad, E. A., Miller, D. D., Flather, C. H., Anderson, E. M., Hupp, J. W. and Anderson, D. R. 1988. Questionable multivariate statistical inference in wildlife habitat and community studies. *J. Wildlife Management* 52: 794–8.

Siegel, S. 1956. *Nonparametric Statistics for the Behavioral Sciences.* McGraw-Hill, New York.

Smith, V. H. 1985. Predictive models for the biomass of blue-green algae in lakes. *Water Resources Bull.* 21: 433–9.

Sokal, R. R. and Rohlf, F. J. 1995. *Biometry.* 3rd Edn. W. H. Freeman, San Francisco, CA.

Underwood, A. J. 1997. *Experiments in Ecology. Their Logical Design and Interpretation Using Analysis of Variance.* Cambridge University Press, Cambridge.

de Vaus, D. A. 1996. *Surveys in Social Research.* 3rd Edn. University College Press, London.

Warner, S. L. 1965. Random response: a survey technique for eliminating evasive answer bias. *J. Am. Stat. Assoc.* 60: 63–9.

Whittacker, R. J. 1987. An application of detrended correspondence analysis and non-metric multidimensional scaling to the identification and analysis of environmental factor complexes and vegetation structures. *J. Ecol.* 75: 363–76.

Wright, J. F., Moss, D., Armitage, P. D. and Furse, M. T. 1984. A preliminary classification of running-water sites in Great Britain based on macro-invertebrate species and the prediction of community type using environmental data. *Freshwat. Biol.* 14: 221–56.

Zullinger, E. M., Ricklefs, R. E., Redford, K. H. and Mace, G. M. 1984. Fitting sigmoidal equations to mammalian growth curves. *J. Mammal.* 65: 607–36.

* Note that several references are given for the original work. In many cases, the results and discussion of these studies can be found in more recent publications.

Further reading

Barnard, C., Gilbert, F. and McGregor, P. 1993. *Asking Questions in Biology*. Addison Wesley Longman, Harlow.

Chase, W. and Bown, F. 1996. *General Statistics*. 3rd Edn. J. Wiley, New York.

Ennos, A. R. and Bailey, S. E. R. 1995. *Environmental Biology*. Addison Wesley Longman, U.K.

Marriott, F. II. C. 1990. *A Dictionary of Statistical Terms*. 5th Edn. Longman, Harlow.

Swinscow, T. D. V. 1990. *Statistics at Square One*. 9th Edn. British Medical Association, London.

Tufte, E. R. 1983. *The Visual Display of Quantitative Information*. Graphics Press Cheshire, Connecticut, USA.

Guide to statistical tests

1. Analysing samples (single variable)

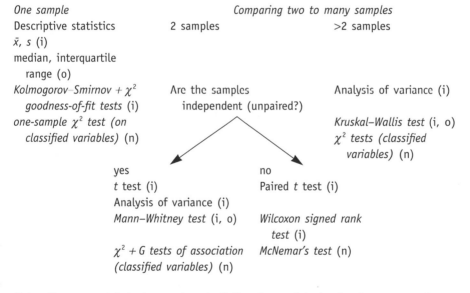

One sample
Descriptive statistics
$\bar{x}$, s (i)
median, interquartile
   range (o)
Kolmogorov–Smirnov + χ^2
   goodness-of-fit tests (i)
one-sample χ^2 test (on
   classified variables) (n)

Comparing two to many samples

2 samples   >2 samples

Are the samples   Analysis of variance (i)
   independent (unpaired?)
                                   Kruskal–Wallis test (i, o)
                                   χ^2 tests (classified
                                      variables) (n)

        yes   no
        t test (i)   Paired *t* test (i)
        Analysis of variance (i)
        Mann–Whitney test (i, o)   *Wilcoxon signed rank*
                                        test (i)
        χ^2 + G tests of association   *McNemar's test (n)*
        (classified variables) (n)

Notes: Non-parametric tests are given in italics. Appropriate levels of measurement are: (n), nominal; (o), ordinal; (i), interval/ratio. Additional tests for normality and homoscedasticity are usually made to validate parametric testing procedures.

2. Relationships between two or more variables (one to many samples)

Two variables
One sample

Many variables
and samples
Multivariate analysis

Estimate one variable from another?

Yes
Regression models 1 and 2 (i)
Non-parametric regression (o,i)

No
Product–moment corrrelation (i)
Spearman's test (o)
ϕ *coefficient of association* (n)

Answers to exercises

6.1 $\chi^2 = 4.204$, significant at $p = 0.05$.

6.2 $\chi^2 = 1.40$, not significant. With Yates' correction, $\chi^2 = 1.10$.

6.3 $\chi^2 = 17.96$, significant at $p = 0.005$. With Yates' correction, $\chi^2 = 16.202$, significant at $p = 0.005$.

6.4 $G_{adj} = 7.7014$, significant at $p = 0.01$.

6.5 $G_{adj} = 11.831$, significant at $p = 0.005$.

6.6 $\chi^2 = 27.89$, significant at $p = 0.005$.

7.1 Approximate normal distribution. Mean = 20.61 °C, standard deviation = 1.811 °C.

7.2 Distribution slightly leptokurtic. Mean = 32.06 ppb, standard deviation = 8.995 ppb.

7.3 Approximate normal distribution. Mean = 15.1 °C, standard deviation = 0.35 °C.

7.4 Approximate normal distribution. Mean = 12.06 ppm, standard deviation = 4.76 ppm.

7.5 $D_{max} = 0.102$, not significant i.e. a fit to the normal distribution. $\bar{x} = 9.195$, $s = 3.133$.

8.1 Variance ratio 1.38, means 12.458, 10.554. $t = 7.55$, significant at $p = 0.001$.

8.2 Variance ratio 1.195, means 21.994, 23.67. $t = 1.805$, not significant as a one-tailed or a two-tailed test.

8.3 City and rural boys, variance ratio 1.44. $t = 3.64$, significant at $p = 0.05$ (one-tailed test). City boys and girls, variance ratio 1.26. $t = 0.49$, not significant (two-tailed test).

8.4 Mann–Whitney $U = 66$, not significant for a two-tailed test.

8.5 Mann–Whitney $U = 209$, yielding $z = 0.764$. Not significant.

8.6 Standard error of the difference $= 0.3037$. $t = 2.414$, significant at $p = 0.05$ one tailed test.

8.7 Standard error of the difference $= 0.0102$. $t = 0.907$, not significant.

8.8 Wilcoxon $T = 6$. Significant at $p = 0.005$ (one-tailed test).

9.1   (i) 13.4 ± 0.42 or $13.0 - 13.8$.
  (ii) 12.1 ± 2.39 or $9.71 - 14.49$.
  (iii) 14.1 ± 0.87 or $13.2 - 15.0$.
  (iv) 5.1 ± 5.91 or -0.8 to 11.0.

9.2   (i) 11.39 ± 2.72.   (ii) 11.39 ± 3.91.

9.3   (i) 0.38 ± 0.095 (or 9.5%).
  (ii) 0.24 ± 0.084 (or 8.4%).
  (iii) 0.31 ± 0.041 (or 4.1%).

10.1 $F = 2.75$, significant at $p = 0.05$.

10.2 $F = 6.08$, significant at $p = 0.01$.

10.3 $F = 68.6$, significant at $p = 0.001$.

10.4 Kruskal–Wallis $H = 5.385$ ($H = 5.386$ corrected for tie). Not significant.

10.5 Kruskal–Wallis $H = 9.25$ (corrected for tie). Significant at $p = 0.05$.

10.6 F (between locations) $= 11.06$, significant at $p = 0.01$. F (between models) $= 5.05$, significant at $p = 0.05$. Total sum of squared deviations 2847.5.

10.7 F (between months) $= 7.87$, significant at $p = 0.01$. F (between heights) not significant.

11.1 $r = -0.561$, significant at $p = 0.05$.

11.2 $r = -0.72$, for log-transformed counts. Significant at $p = 0.001$.

11.3 $r = 0.636$, not significant.

11.4 $r = 0.713$, significant at $p = 0.01$.

11.5 Spearman's $r = 0.605$, significant at $p = 0.01$.

12.1 Regression coefficients $a = 10.38$, $b = 0.017\ 42$. At 1000 m depth, $t = 27.8\ ^{\circ}C$.

12.2 Regression coefficients $a = 16.133$, $b = -4.107 \times 10^{-3}$. At 3000 m altitude the temperature is estimated as $3.812 \pm 1.29\ ^{0}C$.

12.3 Regression coefficients $a = 4.515$, $b = 0.014\ 57$. Minimum height above ground is 145.6 m.

12.4 Regression coefficients $a = 3.465$, $b = 1.5058$. $t = 15.55$, significant at $p = 0.001$ for slope β.

12.5 Regression coefficient with origin forcing $b = 8.561$. Without origin forcing, $a = 2.544$, $b = 4.536$.

Logarithms and antilogarithms

For a positive number (n), the logarithm (log n) is the power to which some other number must be raised to give n

$$\log_b n = x \qquad \text{if } b^x = n$$

The number b is called the *base* of the logarithm. Three bases are in common use, namely base 2, 10 and e (e is the 'natural' base, equal to approximately 2.718). For most calculations the base 10 is used for convenience. For example

$$\log_{10} 100 = 2 \qquad \text{because } 10^2 = 100$$

Before the development of electronic calculators, logarithms to base 10 were used for multiplication and division. The common logarithm ($\log_{10} n$) is often abbreviated to log n and the natural logarithm is abbreviated to ln n. Note that zero and all negative numbers have no logarithms.

In computation, common logarithms are presented in the form of an integer (the *characteristic*) plus a decimal fraction (the *mantissa*). For example 106.9 can be written as 1.069×10^2 and its logarithm is log 1.069 + 2 log 10 = log 1.069 + 2 = 2.028 98 (note the change of multiplier $\times$ to sum +). Electronic calculators perform this calculation (and the antilogarithm) automatically.

An antilogarithm is the number (n) whose logarithm is given. For example, the common antilogarithm of 3.028 98 may be written as

$$10^3 \times \text{antilog}(0.028\ 98)$$

$$n = 1000 \times 1.0690 = 1069.0$$

Statistical tables

Table I Rankits

Rank order	Size of sample = N									
	2	3	4	5	6	7	8	9	10	
1	0.564	0.864	1.029	1.163	1.267	1.352	1.424	1.485	1.539	
2	−0.564	0.000	0.297	0.495	0.642	0.757	0.852	0.932	1.001	
3		−0.864	−0.297	0.000	0.202	0.353	0.473	0.572	0.656	
4			−1.029	−0.495	−0.202	0.000	0.153	0.275	0.376	
5				−1.163	−0.642	−0.353	−0.153	0.000	0.123	
6					−1.267	−0.757	−0.473	−0.275	−0.123	
7						−1.352	−0.852	−0.572	−0.376	
8							−1.424	−0.932	−0.656	
9								−1.485	−1.001	
10									−1.539	

	11	12	13	14	15	16	17	18	19	20
1	1.586	1.629	1.668	1.703	1.736	1.766	1.794	1.820	1.844	1.867
2	1.062	1.116	1.164	1.208	1.248	1.285	1.319	1.350	1.380	1.408
3	0.729	0.793	0.850	0.901	0.948	0.990	1.029	1.066	1.099	1.131
4	0.462	0.537	0.603	0.662	0.715	0.763	0.807	0.848	0.886	0.921
5	0.225	0.312	0.388	0.456	0.516	0.570	0.619	0.665	0.707	0.745
6	0.000	0.103	0.191	0.267	0.335	0.396	0.451	0.502	0.548	0.590
7			0.000	0.088	0.165	0.234	0.295	0.351	0.402	0.448
8					0.000	0.077	0.146	0.208	0.264	0.315
9							0.000	0.069	0.131	0.187
10									0.000	0.062

Table I (Cont'd)

Rank order	Size of sample = N									
	21	22	23	24	25	26	27	28	29	30
1	1.889	1.910	1.929	1.948	1.965	1.982	1.998	2.014	2.029	2.043
2	1.434	1.458	1.481	1.503	1.524	1.544	1.563	1.581	1.599	1.616
3	1.160	1.188	1.214	1.239	1.263	1.285	1.306	1.327	1.346	1.365
4	0.954	0.985	1.014	1.041	1.067	1.091	1.115	1.137	1.158	1.179
5	0.782	0.815	0.847	0.877	0.905	0.932	0.957	0.981	1.004	1.026
6	0.630	0.667	0.701	0.734	0.764	0.793	0.820	0.846	0.871	0.894
7	0.491	0.532	0.569	0.604	0.637	0.668	0.697	0.725	0.752	0.777
8	0.362	0.406	0.446	0.484	0.519	0.553	0.584	0.614	0.642	0.669
9	0.238	0.286	0.330	0.370	0.409	0.444	0.478	0.510	0.540	0.568
10	0.118	0.170	0.218	0.262	0.303	0.341	0.377	0.411	0.443	0.473
11	0.000	0.056	0.108	0.156	0.200	0.241	0.280	0.316	0.350	0.382
12			0.000	0.052	0.100	0.144	0.185	0.224	0.260	0.294
13					0.000	0.048	0.092	0.134	0.172	0.209
14							0.000	0.044	0.086	0.125
15									0.000	0.041

Table II Areas under the standard normal curve 191

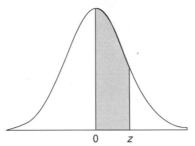

z	0.00	0.01	0.02	0.03	0.04	0.05	0.06	0.07	0.08	0.09
0.0	0.0000	0.0040	0.0080	0.0120	0.0160	0.0199	0.0239	0.0279	0.0319	0.0359
0.1	0.0398	0.0438	0.0478	0.0517	0.0557	0.0596	0.0636	0.0675	0.0714	0.0753
0.2	0.0793	0.0832	0.0871	0.0910	0.0948	0.0987	0.1026	0.1064	0.1103	0.1141
0.3	0.1179	0.1217	0.1255	0.1293	0.1331	0.1368	0.1406	0.1443	0.1480	0.1517
0.4	0.1554	0.1591	0.1628	0.1664	0.1700	0.1736	0.1772	0.1808	0.1844	0.1879
0.5	0.1915	0.1950	0.1985	0.2019	0.2054	0.2088	0.2123	0.2157	0.2190	0.2224
0.6	0.2257	0.2291	0.2324	0.2357	0.2389	0.2422	0.2454	0.2486	0.2517	0.2549
0.7	0.2580	0.2611	0.2642	0.2673	0.2704	0.2734	0.2764	0.2794	0.2823	0.2852
0.8	0.2881	0.2910	0.2939	0.2967	0.2995	0.3023	0.3051	0.3078	0.3106	0.3133
0.9	0.3159	0.3186	0.3212	0.3238	0.3264	0.3289	0.3315	0.3340	0.3365	0.3389
1.0	0.3413	0.3438	0.3461	0.3485	0.3508	0.3531	0.3554	0.3577	0.3599	0.3621
1.1	0.3643	0.3665	0.3686	0.3708	0.3729	0.3749	0.3770	0.3790	0.3810	0.3830
1.2	0.3849	0.3869	0.3888	0.3907	0.3925	0.3944	0.3962	0.3980	0.3997	0.4015
1.3	0.4032	0.4049	0.4066	0.4082	0.4099	0.4115	0.4131	0.4147	0.4162	0.4177
1.4	0.4192	0.4207	0.4222	0.4236	0.4251	0.4265	0.4279	0.4292	0.4306	0.4319
1.5	0.4332	0.4345	0.4357	0.4370	0.4382	0.4394	0.4406	0.4418	0.4429	0.4441
1.6	0.4452	0.4463	0.4474	0.4484	0.4495	0.4505	0.4515	0.4525	0.4535	0.4545
1.7	0.4554	0.4564	0.4573	0.4582	0.4591	0.4599	0.4608	0.4616	0.4625	0.4633
1.8	0.4641	0.4649	0.4656	0.4664	0.4671	0.4678	0.4686	0.4693	0.4699	0.4706
1.9	0.4713	0.4719	0.4726	0.4732	0.4738	0.4744	0.4750	0.4756	0.4761	0.4767
2.0	0.4772	0.4778	0.4783	0.4788	0.4793	0.4798	0.4803	0.4808	0.4812	0.4817
2.1	0.4821	0.4826	0.4830	0.4834	0.4838	0.4842	0.4846	0.4850	0.4854	0.4857
2.2	0.4861	0.4864	0.4868	0.4871	0.4875	0.4878	0.4881	0.4884	0.4887	0.4890
2.3	0.4893	0.4896	0.4898	0.4901	0.4904	0.4906	0.4909	0.4911	0.4913	0.4916
2.4	0.4918	0.4920	0.4922	0.4925	0.4927	0.4929	0.4931	0.4932	0.4934	0.4936
2.5	0.4938	0.4940	0.4941	0.4943	0.4945	0.4946	0.4948	0.4949	0.4951	0.4952
2.6	0.4953	0.4955	0.4956	0.4957	0.4959	0.4960	0.4961	0.4962	0.4963	0.4964
2.7	0.4965	0.4966	0.4967	0.4968	0.4969	0.4970	0.4971	0.4972	0.4973	0.4974
2.8	0.4974	0.4975	0.4976	0.4977	0.4977	0.4978	0.4979	0.4979	0.4980	0.4981
2.9	0.4981	0.4982	0.4982	0.4983	0.4984	0.4984	0.4985	0.4985	0.4986	0.4986
3.0	0.4987	0.4987	0.4987	0.4988	0.4988	0.4989	0.4989	0.4989	0.4990	0.4990
3.1	0.4990	0.4991	0.4991	0.4991	0.4992	0.4992	0.4992	0.4992	0.4993	0.4993
3.2	0.4993	0.4993	0.4994	0.4994	0.4994	0.4994	0.4994	0.4995	0.4995	0.4995
3.3	0.4995	0.4995	0.4995	0.4996	0.4996	0.4996	0.4996	0.4996	0.4996	0.4997
3.4	0.4997	0.4997	0.4997	0.4997	0.4997	0.4997	0.4997	0.4997	0.4997	0.4998

Table III Values of the t statistic

	Level of significance for one-tailed test					
	0.10	0.05	0.025	0.01	0.005	0.0005
df	Level of significance for two-tailed test					
	0.20	0.10	0.05	0.02	0.01	0.001
1	3.078	6.314	12.706	31.821	63.657	636.619
2	1.886	2.920	4.303	6.965	9.925	31.598
3	1.638	2.353	3.182	4.541	5.841	12.941
4	1.533	2.132	2.776	3.747	4.604	8.610
5	1.476	2.015	2.571	3.365	4.032	6.859
6	1.440	1.943	2.447	3.143	3.707	5.959
7	1.415	1.895	2.365	2.998	3.499	5.405
8	1.397	1.860	2.306	2.896	3.355	5.041
9	1.383	1.833	2.262	2.821	3.250	4.781
10	1.372	1.812	2.228	2.764	3.169	4.587
11	1.363	1.796	2.201	2.718	3.106	4.437
12	1.356	1.782	2.179	2.681	3.055	4.318
13	1.350	1.771	2.160	2.650	3.012	4.221
14	1.345	1.761	2.145	2.624	2.977	4.140
15	1.341	1.753	2.131	2.602	2.947	4.073
16	1.337	1.746	2.120	2.583	2.921	4.015
17	1.333	1.740	2.110	2.567	2.898	3.965
18	1.330	1.734	2.101	2.552	2.878	3.922
19	1.328	1.729	2.093	2.539	2.861	3.883
20	1.325	1.725	2.086	2.528	2.845	3.850
21	1.323	1.721	2.080	2.518	2.831	3.819
22	1.321	1.717	2.074	2.508	2.819	3.792
23	1.319	1.714	2.069	2.500	2.807	3.767
24	1.318	1.711	2.064	2.492	2.797	3.745
25	1.316	1.708	2.060	2.485	2.787	3.725
26	1.315	1.706	2.056	2.479	2.779	3.707
27	1.314	1.703	2.052	2.473	2.771	3.690
28	1.313	1.701	2.048	2.467	2.763	3.674
29	1.311	1.699	2.045	2.462	2.756	3.659
30	1.310	1.697	2.042	2.457	2.750	3.646
40	1.303	1.684	2.021	2.423	2.704	3.551
60	1.296	1.671	2.000	2.390	2.660	3.460
120	1.289	1.658	1.980	2.358	2.617	3.373
∞	**1.282**	**1.645**	**1.960**	**2.326**	**2.576**	**3.291**

Table IV The Mann–Whitney U statistic

Critical values of U for a one-tailed test at $\alpha = 0.001$ or for a two-tailed test at $\alpha = 0.002$

n_1 \ n_2	9	10	11	12	13	14	15	16	17	18	19	20
1												
2												
3									0	0	0	0
4		0	0	0	1	1	1	2	2	3	3	3
5	1	1	2	2	3	3	4	5	5	6	7	7
6	2	3	4	4	5	6	7	8	9	10	11	12
7	3	5	6	7	8	9	10	11	13	14	15	16
8	5	6	8	9	11	12	14	15	17	18	20	21
9	7	8	10	12	14	15	17	19	21	23	25	26
10	8	10	12	14	17	19	21	23	25	27	29	32
11	10	12	15	17	20	22	24	27	29	32	34	37
12	12	14	17	20	23	25	28	31	34	37	40	42
13	14	17	20	23	26	29	32	35	38	42	45	48
14	15	19	22	25	29	32	36	39	43	46	50	54
15	17	21	24	28	32	36	40	43	47	51	55	59
16	19	23	27	31	35	39	43	48	52	56	60	65
17	21	25	29	34	38	43	47	52	57	61	66	70
18	23	27	32	37	42	46	51	56	61	66	71	76
19	25	29	34	40	45	50	55	60	66	71	77	82
20	26	32	37	42	48	54	59	65	70	76	82	88

Table IV (Cont'd)

Critical values of U for a one-tailed test at $\alpha = 0.025$ or for a two-tailed test at $\alpha = 0.05$

n_1 \ n_2	9	10	11	12	13	14	15	16	17	18	19	20
1												
2	0	0	0	1	1	1	1	1	2	2	2	2
3	2	3	3	4	4	5	5	6	6	7	7	8
4	4	5	6	7	8	9	10	11	11	12	13	13
5	7	8	9	11	12	13	14	15	17	18	19	20
6	10	11	13	14	16	17	19	21	22	24	25	27
7	12	14	16	18	20	22	24	26	28	30	32	34
8	15	17	19	22	24	26	29	31	34	36	38	41
9	17	20	23	26	28	31	34	37	39	42	45	48
10	20	23	26	29	33	36	39	42	45	48	52	55
11	23	26	30	33	37	40	44	47	51	55	58	62
12	26	29	33	37	41	45	49	53	57	61	65	69
13	28	33	37	41	45	50	54	59	63	67	72	76
14	31	36	40	45	50	55	59	64	67	74	78	83
15	34	39	44	49	54	59	64	70	75	80	85	90
16	37	42	47	53	59	64	70	75	81	86	92	98
17	39	45	51	57	63	67	75	81	87	93	99	105
18	42	48	55	61	67	74	80	86	93	99	106	112
19	45	52	58	65	72	78	85	92	99	106	113	119
20	48	55	62	69	76	83	90	98	105	112	119	127

Table V Critical values of T in the Wilcoxon test

	Level of significance for one-tailed test	
	0.025	*0.005*
	Level of significance for two-tailed test	
n	*0.05*	*0.01*
6	0	–
7	2	–
8	4	0
9	6	2
10	8	3
11	11	5
12	14	7
13	17	10
14	21	13
15	25	16
16	30	20
17	35	23
18	40	28
19	46	32
20	52	38
21	59	43
22	66	49
23	73	55
24	81	61
25	89	68

Table VI Values of the χ^2 distribution for v degrees of freedom and probability level P per cent

P	5%	1%	0.1%
$v = 1$	3.84	6.63	10.83
2	5.99	9.21	13.81
3	7.81	11.34	16.27
4	9.49	13.28	18.47
5	11.07	15.09	20.52
6	12.59	16.81	22.46
7	14.07	18.48	24.32
8	15.51	20.09	26.12
9	16.92	21.67	27.88
10	18.31	23.21	29.59
11	19.68	24.73	31.26
12	21.03	26.22	32.91
13	22.36	27.69	34.53
14	23.68	29.14	36.12
15	25.00	30.58	37.70
16	26.30	32.00	39.25
17	27.59	33.41	40.79
18	28.87	34.81	42.31
19	30.14	36.19	43.82
20	31.41	37.57	45.31
21	32.67	38.93	46.80
22	33.92	40.29	48.27
23	35.17	41.64	49.73
24	36.42	42.98	51.18
25	37.65	44.31	52.62
26	38.89	45.64	54.05
27	40.11	46.96	55.48
28	41.34	48.28	56.89
29	42.56	49.59	58.30
30	43.77	50.89	59.70
40	55.76	63.69	73.40
50	67.50	76.15	86.66
60	79.08	88.38	99.61
70	90.53	100.4	112.3
80	101.9	112.3	124.8
90	113.1	124.1	137.2
100	124.3	135.8	149.4

Table VII Critical values of the Kolmogorov–Smirnov statistic for a one-sample test

Sample size n	Significance level	
	5%	*1%*
1	0.975	0.995
2	0.842	0.929
3	0.708	0.828
4	0.624	0.733
5	0.565	0.669
6	0.521	0.618
7	0.486	0.577
8	0.457	0.543
9	0.432	0.514
10	0.410	0.490
11	0.391	0.468
12	0.375	0.450
13	0.361	0.433
14	0.349	0.418
15	0.338	0.404
16	0.328	0.392
17	0.318	0.381
18	0.309	0.371
19	0.301	0.363
20	0.294	0.356
25	0.27	0.32
30	0.24	0.29
35	0.23	0.27
∞	1.36	1.63

Table VIII

1 per cent points of the F distribution

$v_1 =$	1	2	3	4	5	6	7	8	10	12	24	∞
$v_2 = 1$	4052	5000	5403	5625	5764	5859	5928	5981	6056	6106	6235	6366
2	98.5	99.0	99.2	99.2	99.3	99.3	99.4	99.4	99.4	99.4	99.5	99.5
3	34.1	30.8	29.5	28.7	28.2	27.9	27.7	27.5	27.2	27.1	26.6	26.1
4	21.2	18.0	16.7	16.0	15.5	15.2	15.0	14.8	14.5	14.4	13.9	13.5
5	16.26	13.27	12.06	11.39	10.97	10.67	10.46	10.29	10.05	9.89	9.47	9.02
6	13.74	10.92	9.78	9.15	8.75	8.47	8.26	8.10	7.87	7.72	7.31	6.88
7	12.25	9.55	8.45	7.85	7.46	7.19	6.99	6.84	6.62	6.47	6.07	5.65
8	11.26	8.65	7.59	7.01	6.63	6.37	6.18	6.03	5.81	5.67	5.28	4.86
9	10.56	8.02	6.99	6.42	6.06	5.80	5.61	5.47	5.26	5.11	4.73	4.31
10	10.04	7.56	6.55	5.99	5.64	5.39	5.20	5.06	4.85	4.71	4.33	3.91
11	9.65	7.21	6.22	5.67	5.32	5.07	4.89	4.74	4.54	4.40	4.02	3.60
12	9.33	6.93	5.95	5.41	5.06	4.82	4.64	4.50	4.30	4.16	3.78	3.36
13	9.07	6.70	5.74	5.21	4.86	4.62	4.44	4.30	4.10	3.96	3.59	3.17
14	8.86	6.51	5.56	5.04	4.70	4.46	4.28	4.14	3.94	3.80	3.43	3.00
15	8.68	6.36	5.42	4.89	4.56	4.32	4.14	4.00	3.80	3.67	3.29	2.87
16	8.53	6.23	5.29	4.77	4.44	4.20	4.03	3.89	3.69	3.55	3.18	2.75
17	8.40	6.11	5.18	4.67	4.34	4.10	3.93	3.79	3.59	3.46	3.08	2.65
18	8.29	6.01	5.09	4.58	4.25	4.01	3.84	3.71	3.51	3.37	3.00	2.57
19	8.18	5.93	5.01	4.50	4.17	3.94	3.77	3.63	3.43	3.30	2.92	2.49

20	8.10	5.85	4.94	4.43	4.10	3.87	3.70	3.56	3.37	3.23	2.86	2.42
21	8.02	5.78	4.87	4.37	4.04	3.81	3.64	3.51	3.31	3.17	2.80	2.36
22	7.95	5.72	4.82	4.31	3.99	3.76	3.59	3.45	3.26	3.12	2.75	2.31
23	7.88	5.66	4.76	4.26	3.94	3.71	3.54	3.41	3.21	3.07	2.70	2.26
24	7.82	5.61	4.72	4.22	3.90	3.67	3.50	3.36	3.17	3.03	2.66	2.21
25	7.77	5.57	4.68	4.18	3.86	3.63	3.46	3.32	3.13	2.99	2.62	2.17
26	7.72	5.53	4.64	4.14	3.82	3.59	3.42	3.29	3.09	2.96	2.58	2.13
27	7.68	5.49	4.60	4.11	3.78	3.56	3.39	3.26	3.06	2.93	2.55	2.10
28	7.64	5.45	4.57	4.07	3.75	3.53	3.36	3.23	3.03	2.90	2.52	2.06
29	7.60	5.42	4.54	4.04	3.73	3.50	3.33	3.20	3.00	2.87	2.49	2.03
30	7.56	5.39	4.51	4.02	3.70	3.47	3.30	3.17	2.98	2.84	2.47	2.01
32	7.50	5.34	4.46	3.97	3.65	3.43	3.26	3.13	2.93	2.80	2.42	1.96
34	7.45	5.29	4.42	3.93	3.61	3.39	3.22	3.09	2.90	2.76	2.38	1.91
36	7.40	5.25	4.38	3.89	3.58	3.35	3.18	3.05	2.86	2.72	2.35	1.87
38	7.35	5.21	4.34	3.86	3.54	3.32	3.15	3.02	2.83	2.69	2.32	1.84
40	7.31	5.18	4.31	3.83	3.51	3.29	3.12	2.99	2.80	2.66	2.29	1.80
60	7.08	4.98	4.13	3.65	3.34	3.12	2.95	2.82	2.63	2.50	2.12	1.60
120	6.85	4.79	3.95	3.48	3.17	2.96	2.79	2.66	2.47	2.34	1.95	1.38
∞	6.63	4.61	3.78	3.32	3.02	2.80	2.64	2.51	2.32	2.18	1.79	1.00

Table VIII (Cont'd)

5 per cent points of the F distribution

$v_1 =$	1	2	3	4	5	6	7	8	10	12	24	∞
$v_2 = 1$	161.4	199.5	215.7	224.6	230.2	234.0	236.8	238.9	241.9	243.9	249.0	254.3
2	18.5	19.0	19.2	19.2	19.3	19.3	19.4	19.4	19.4	19.4	19.5	19.5
3	10.13	9.55	9.28	9.12	9.01	8.94	8.89	8.85	8.79	8.74	8.64	8.53
4	7.71	6.94	6.59	6.39	6.26	6.16	6.09	6.04	5.96	5.91	5.77	5.63
5	6.61	5.79	5.41	5.19	5.05	4.95	4.88	4.82	4.74	4.68	4.53	4.36
6	5.99	5.14	4.76	4.53	4.39	4.28	4.21	4.15	4.06	4.00	3.84	3.67
7	5.59	4.74	4.35	4.12	3.97	3.87	3.79	3.73	3.64	3.57	3.41	3.23
8	5.32	4.46	4.07	3.84	3.69	3.58	3.50	3.44	3.35	3.28	3.12	2.93
9	5.12	4.26	3.86	3.63	3.48	3.37	3.29	3.23	3.14	3.07	2.90	2.71
10	4.96	4.10	3.71	3.48	3.33	3.22	3.14	3.07	2.98	2.91	2.74	2.54
11	4.84	3.98	3.59	3.36	3.20	3.09	3.01	2.95	2.85	2.79	2.61	2.40
12	4.75	3.89	3.49	3.26	3.11	3.00	2.91	2.85	2.75	2.69	2.51	2.30
13	4.67	3.81	3.41	3.18	3.03	2.92	2.83	2.77	2.67	2.60	2.42	2.21
14	4.60	3.74	3.34	3.11	2.96	2.85	2.76	2.70	2.60	2.53	2.35	2.13
15	4.54	3.68	3.29	3.06	2.90	2.79	2.71	2.64	2.54	2.48	2.29	2.07
16	4.49	3.63	3.24	3.01	2.85	2.74	2.66	2.59	2.49	2.42	2.24	2.01
17	4.45	3.59	3.20	2.96	2.81	2.70	2.61	2.55	2.45	2.38	2.19	1.96
18	4.41	3.55	3.16	2.93	2.77	2.66	2.58	2.51	2.41	2.34	2.15	1.92
19	4.38	3.52	3.13	2.90	2.74	2.63	2.54	2.48	2.38	2.31	2.11	1.88

20	4.35	3.49	3.10	2.87	2.71	2.60	2.51	2.45	2.35	2.28	2.08	1.84
21	4.32	3.47	3.07	2.84	2.68	2.57	2.49	2.42	2.32	2.25	2.05	1.81
22	4.30	3.44	3.05	2.82	2.66	2.55	2.46	2.40	2.30	2.23	2.03	1.78
23	4.28	3.42	3.03	2.80	2.64	2.53	2.44	2.37	2.27	2.20	2.00	1.76
24	4.26	3.40	3.01	2.78	2.62	2.51	2.42	2.36	2.25	2.18	1.98	1.73
25	4.24	3.39	2.99	2.76	2.60	2.49	2.40	2.34	2.24	2.16	1.96	1.71
26	4.23	3.37	2.98	2.74	2.59	2.47	2.39	2.32	2.22	2.15	1.95	1.69
27	4.21	3.35	2.96	2.73	2.57	2.46	2.37	2.31	2.20	2.13	1.93	1.67
28	4.20	3.34	2.95	2.71	2.56	2.45	2.36	2.29	2.19	2.12	1.91	1.65
29	4.18	3.33	2.93	2.70	2.55	2.43	2.35	2.28	2.18	2.10	1.90	1.64
30	4.17	3.32	2.92	2.69	2.53	2.42	2.33	2.27	2.16	2.09	1.89	1.62
32	4.15	3.29	2.90	2.67	2.51	2.40	2.31	2.24	2.14	2.07	1.86	1.59
34	4.13	3.28	2.88	2.65	2.49	2.38	2.29	2.23	2.12	2.05	1.84	1.57
36	4.11	3.26	2.87	2.63	2.48	2.36	2.28	2.21	2.11	2.03	1.82	1.55
38	4.10	3.24	2.85	2.62	2.46	2.35	2.26	2.19	2.09	2.02	1.81	1.53
40	4.08	3.23	2.84	2.61	2.45	2.34	2.25	2.18	2.08	2.00	1.79	1.51
60	4.00	3.15	2.76	2.53	2.37	2.25	2.17	2.10	1.99	1.92	1.70	1.39
120	3.92	3.07	2.68	2.45	2.29	2.18	2.09	2.02	1.91	1.83	1.61	1.25
∞	3.84	3.00	2.60	2.37	2.21	2.10	2.01	1.94	1.83	1.75	1.52	1.00

Table IX Values of q used in the Tukey test

df	a = Number of samples being compared								
	2	3	4	5	6	7	8	9	10
1	17.97	26.98	32.82	37.08	40.41	43.12	45.40	47.36	49.07
2	6.08	8.33	9.80	10.88	11.74	12.44	13.03	13.54	13.99
3	4.50	5.91	6.82	7.50	8.04	8.48	8.85	9.18	9.46
4	3.93	5.04	5.76	6.29	6.71	7.05	7.35	7.60	7.83
5	3.64	4.60	5.22	5.67	6.03	6.33	6.58	6.80	6.99
6	3.46	4.34	4.90	5.30	5.63	5.90	6.12	6.32	6.49
7	3.34	4.16	4.68	5.06	5.36	5.61	5.82	6.00	6.16
8	3.26	4.04	4.53	4.89	5.17	5.40	5.60	5.77	5.92
9	3.20	3.95	4.41	4.76	5.02	5.24	5.43	5.59	5.74
10	3.15	3.88	4.33	4.65	4.91	5.12	5.30	5.46	5.60
11	3.11	3.82	4.26	4.57	4.82	5.03	5.20	5.35	5.49
12	3.08	3.77	4.20	4.51	4.75	4.95	5.12	5.27	5.39
13	3.06	3.73	4.15	4.45	4.69	4.88	5.05	5.19	5.32
14	3.03	3.70	4.11	4.41	4.64	4.83	4.99	5.13	5.25
15	3.01	3.67	4.08	4.37	4.59	4.78	4.94	5.08	5.20
16	3.00	3.65	4.05	4.33	4.56	4.74	4.90	5.03	5.15
17	2.98	3.63	4.02	4.30	4.52	4.70	4.86	4.99	5.11
18	2.97	3.61	4.00	4.28	4.49	4.67	4.82	4.96	5.07
19	2.96	3.59	3.98	4.25	4.47	4.65	4.79	4.92	5.04
20	2.95	3.58	3.96	4.23	4.45	4.62	4.77	4.90	5.01
24	2.92	3.53	3.90	4.17	4.37	4.54	4.68	4.81	4.92
30	2.89	3.49	3.85	4.10	4.30	4.46	4.60	4.72	4.82
40	2.86	3.44	3.79	4.04	4.23	4.39	4.52	4.63	4.73
60	2.83	3.40	3.74	3.98	4.16	4.31	4.44	4.55	4.65
120	2.80	3.36	3.68	3.92	4.10	4.24	4.36	4.47	4.56

Table IX (Cont'd)

203

df	\(a\) = Number of samples being compared									
	11	12	13	14	15	16	17	18	19	20
1	50.59	51.96	53.20	54.33	55.36	56.32	57.22	58.04	58.83	59.56
2	14.39	14.75	15.08	15.38	15.65	15.91	16.14	16.37	16.57	16.77
3	9.72	9.95	10.15	10.35	10.52	10.69	10.84	10.98	11.11	11.24
4	8.03	8.21	8.37	8.52	8.66	8.79	8.91	9.03	9.13	9.23
5	7.17	7.32	7.47	7.60	7.72	7.83	7.93	8.03	8.12	8.21
6	6.65	6.79	6.92	7.03	7.14	7.24	7.34	7.43	7.51	7.59
7	6.30	6.43	6.55	6.66	6.76	6.85	6.94	7.02	7.10	7.17
8	6.05	6.18	6.29	6.39	6.48	6.57	6.65	6.73	6.80	6.87
9	5.87	5.98	6.09	6.19	6.28	6.36	6.44	6.51	6.58	6.64
10	5.72	5.83	5.93	6.03	6.11	6.19	6.27	6.34	6.40	6.47
11	5.61	5.71	5.81	5.90	5.98	6.06	6.13	6.20	6.27	6.33
12	5.51	5.61	5.71	5.80	5.88	5.95	6.02	6.09	6.15	6.21
13	5.43	5.53	5.63	5.71	5.79	5.86	5.93	5.99	6.05	6.11
14	5.36	5.46	5.55	5.64	5.71	5.79	5.85	5.91	5.97	6.03
15	5.31	5.40	5.49	5.57	5.65	5.72	5.78	5.85	5.90	5.96
16	5.26	5.35	5.44	5.52	5.59	5.66	5.73	5.79	5.84	5.90
17	5.21	5.31	5.39	5.47	5.54	5.61	5.67	5.73	5.79	5.84
18	5.17	5.27	5.35	5.43	5.50	5.57	5.63	5.69	5.74	5.79
19	5.14	5.23	5.31	5.39	5.46	5.53	5.59	5.65	5.70	5.75
20	5.11	5.20	5.28	5.36	5.43	5.49	5.55	5.61	5.66	5.71
24	5.01	5.10	5.18	5.25	5.32	5.38	5.44	5.49	5.55	5.59
30	4.92	5.00	5.08	5.15	5.21	5.27	5.33	5.38	5.43	5.47
40	4.82	4.90	4.98	5.04	5.11	5.16	5.22	5.27	5.31	5.36
60	4.73	4.81	4.88	4.94	5.00	5.06	5.11	5.15	5.20	5.24
120	4.64	4.71	4.78	4.84	4.90	4.95	5.00	5.04	5.09	5.13

Table X Probabilities associated with values of H in the Kruskal–Wallis test

Sample sizes					Sample sizes				
n_1	n_2	n_3	H	p	n_1	n_2	n_3	H	p
2	1	1	2.7000	0.500	4	3	2	6.4444	0.008
2	2	1	3.6000	0.200				6.3000	0.011
								5.4444	0.046
2	2	2	4.5714	0.067				5.4000	0.051
			3.7143	0.200				4.5111	0.098
								4.4444	0.102
3	1	1	3.2000	0.300					
					4	3	3	6.7455	0.010
3	2	1	4.2857	0.100				6.7091	0.013
			3.8571	0.133				5.7909	0.046
								5.7273	0.050
3	2	2	5.3572	0.029				4.7091	0.092
			4.7143	0.048				4.7000	0.101
			4.5000	0.067					
			4.4643	0.105	4	4	1	6.6667	0.010
								6.1667	0.022
3	3	1	5.1429	0.043				4.9667	0.048
			4.5714	0.100				4.8667	0.054
			4.0000	0.129				4.1667	0.082
								4.0667	0.102
3	3	2	6.2500	0.011					
			5.3611	0.032					
			5.1389	0.061	4	4	2	7.0364	0.006
			4.5556	0.100				6.8727	0.011
			4.2500	0.121				5.4545	0.046
								5.2364	0.052
3	3	3	7.2000	0.004				4.5545	0.098
			6.4889	0.011				4.4455	0.103
			5.6889	0.029					
			5.6000	0.050	4	4	3	7.1439	0.010
			5.0667	0.086				7.1364	0.011
			4.6222	0.100				5.5985	0.049
								5.5758	0.051
4	1	1	3.5714	0.200				4.5455	0.099
								4.4773	0.102
4	2	1	4.8214	0.057					
			4.5000	0.076	4	4	4	7.6538	0.008
			4.0179	0.114				7.5385	0.011
								5.6923	0.049
4	2	2	6.0000	0.014				5.6538	0.054
			5.3333	0.033				4.6539	0.097
			5.1250	0.052				4.5001	0.104
			4.4583	0.100					
			4.1667	0.105	5	1	1	3.8571	0.143
4	3	1	5.8333	0.021	5	2	1	5.2500	0.036
			5.2083	0.050				5.0000	0.018
			5.0000	0.057				4.4500	0.071
			4.0556	0.093				4.2000	0.095
			3.8889	0.129				4.0500	0.119

Table X (Cont'd)

205

Sample sizes					Sample sizes				
n_1	n_2	n_3	H	p	n_1	n_2	n_3	H	p
5	2	2	6.5333	0.008	5	4	4	7.7604	0.009
			6.1333	0.013				7.7440	0.011
			5.1600	0.034				5.6571	0.049
			5.0400	0.056				5.6176	0.050
			4.3733	0.090				4.6187	0.100
			4.2933	0.122				4.5527	0.102
5	3	1	6.4000	0.012	5	5	1	7.3091	0.009
			4.9600	0.048				6.8364	0.011
			4.8711	0.052				5.1273	0.046
			4.0178	0.095				4.9091	0.053
			3.8400	0.123				4.1091	0.086
								4.0364	0.105
5	3	2	6.9091	0.009	5	5	2	7.3385	0.010
			6.8218	0.010				7.2692	0.010
			5.2509	0.049				5.3385	0.047
			5.1055	0.052				5.2462	0.051
			4.6509	0.091				4.6231	0.097
			4.4945	0.101				4.5077	0.100
5	3	3	7.0788	0.009	5	5	3	7.5780	0.010
			6.9818	0.011				7.5429	0.010
			5.6485	0.049				5.7055	0.046
			5.5152	0.051				5.6264	0.051
			4.5333	0.097				4.5451	0.100
			4.4121	0.109				4.5363	0.102
5	4	1	6.9545	0.008	5	5	4	7.8229	0.010
			6.8400	0.011				7.7914	0.010
			4.9855	0.044				5.6657	0.049
			4.8600	0.056				5.6429	0.050
			3.9873	0.098				4.5229	0.099
			3.9600	0.102				4.5200	0.101
5	4	2	7.2045	0.009	5	5	5	8.0000	0.009
			7.1182	0.010				7.9800	0.010
			5.2727	0.049				5.7800	0.049
			5.2682	0.050				5.6600	0.051
			4.5409	0.098				4.5600	0.100
			4.5182	0.101				4.5000	0.102
5	4	3	7.4449	0.010					
			7.3949	0.011					
			5.6564	0.019					
			5.6308	0.050					
			4.5487	0.099					
			4.5231	0.103					

Table XI Product–moment correlation coefficients at the 5% and 1% levels of significance

Degrees of freedom	5%	1%	Degrees of freedom	5%	1%
1	0.997	1.000	24	0.388	0.496
2	0.950	0.990	25	0.381	0.487
3	0.878	0.959	26	0.374	0.478
4	0.811	0.917	27	0.367	0.470
5	0.754	0.874	28	0.361	0.463
6	0.707	0.834	29	0.355	0.456
7	0.666	0.798	30	0.349	0.449
8	0.632	0.765	35	0.325	0.418
9	0.602	0.735	40	0.304	0.393
10	0.576	0.708	45	0.288	0.372
11	0.553	0.684	50	0.273	0.354
12	0.532	0.661	60	0.250	0.325
13	0.514	0.641	70	0.232	0.302
14	0.497	0.623	80	0.217	0.283
15	0.482	0.606	90	0.205	0.267
16	0.468	0.590	100	0.195	0.254
17	0.456	0.575	125	0.174	0.228
18	0.444	0.561	150	0.159	0.208
19	0.433	0.549	200	0.138	0.181
20	0.423	0.537	300	0.113	0.148
21	0.413	0.526	400	0.098	0.128
22	0.404	0.515	500	0.088	0.115
23	0.396	0.505	1000	0.062	0.081

Table XII Critical values of Spearman's correlation coefficient

207

	One-tailed test					
	0.100	0.050	0.025	0.010	0.005	0.001
	Two-tailed test					
n	0.200	0.100	0.050	0.020	0.010	0.002
4	1.000	1.000				
5	0.800	0.900	1.000	1.000		
6	0.657	0.829	0.886	0.943	1.000	
7	0.571	0.714	0.786	0.893	0.929	1.000
8	0.524	0.643	0.738	0.833	0.881	0.952
9	0.483	0.600	0.700	0.783	0.833	0.917
10	0.455	0.564	0.648	0.745	0.794	0.879
11	0.427	0.536	0.618	0.709	0.755	0.845
12	0.406	0.503	0.587	0.678	0.727	0.818
13	0.385	0.484	0.560	0.648	0.703	0.791
14	0.367	0.464	0.538	0.626	0.679	0.771
15	0.354	0.446	0.521	0.604	0.657	0.750
16	0.341	0.429	0.503	0.585	0.635	0.729
17	0.328	0.414	0.488	0.566	0.618	0.711
18	0.317	0.401	0.474	0.550	0.600	0.692
19	0.309	0.391	0.460	0.535	0.584	0.675
20	0.299	0.380	0.447	0.522	0.570	0.660
21	0.292	0.370	0.436	0.509	0.556	0.647
22	0.284	0.361	0.425	0.497	0.544	0.633
23	0.278	0.353	0.416	0.486	0.532	0.620
24	0.271	0.344	0.407	0.476	0.521	0.608
25	0.265	0.337	0.398	0.466	0.511	0.597
26	0.259	0.331	0.390	0.457	0.501	0.586
27	0.255	0.324	0.383	0.449	0.492	0.576
28	0.250	0.318	0.375	0.441	0.483	0.567
29	0.245	0.312	0.369	0.433	0.475	0.557
30	0.240	0.306	0.362	0.426	0.467	0.548

Glossary

abscissa The x coordinate or x axis measured along the page from the origin.

accuracy The closeness of a measurement to its true value.

bar chart Graphical display of frequency or magnitude using rectangles or bars. Discontinuous variables rather than continuous variables are normally displayed along the x axis.

centroid The point on a scattergraph whose abscissa and ordinate correspond to the means of the plotted variates.

class interval Refers to the upper and lower limits of a class into which a variate is grouped in the construction of a frequency histogram.

continuous variable A variable which can take any value within the set.

correlation The relationship between measurable variates or ranks; it is most frequently sought between two variates. In a wider sense it includes the association between attributes in cross-tabulations.

cross-tabulation A table used to classify members according to qualitative characteristics, the simplest being a two-way or contingency table.

datum A single measurement or piece of information. Plural *data* (often used as singular in the USA).

degrees of freedom A form of information tax relating to the number (n) of observations in a sample. For statistical calculations, degrees of freedom are normally obtained by subtracting integers from n (e.g. $n-1$) to maintain independence of the information.

dendrogram A tree diagram used in cluster analysis to display the relationships between samples or individuals.

denominator The divisor in a fraction, e.g. in 7/9 the number 9.

dependent In regression where the value of a variate y depends on the value(s) of variate(s) x.

deviation The difference between one measurement and another. Usually one of these measurements is a statistic, such as the mean. Often the deviation is obtained by simple subtraction, but this is not always the case (cf. standard deviation).

discontinuous A variable which can take only specified values, such as integers (as in counts of pollen grains).

dispersion The degree of scatter shown by observations.

exponent A number placed as a superscipt to another number or variable to indicate repeated multiplication, e.g. 3 in 10^3 ($10 \times 10 \times 10$).

frequency The number of occurrences of a particular value or event falling into a specified class.

heteroscedastic Where the variance of one variate differs for fixed values of another variate.

homoscedastic Where the variance of one variate is the same for fixed values of another variate.

independent In regression where a variable x is measured without error, and is used to obtain a value of the dependent variable.

intercept The distance from the origin (i.e. the ordinate) where a regression line cuts the y axis.

interference Used when the combined effect of two different treatments is less than the sum of the effects of the individual treatments.

interpolation Where two variables are related by a curve drawn on a graph (or tabulated curves), a method for estimating an unknown value of one variable from two values of the other variable.

interquartile The variate distance between the upper and lower quartiles, encompassing one-half of the total frequency.

interval A scale of measurement amenable to arithmetic procedures but with no absolute zero point (cf. ratio).

location The notion of centrality when applied to a collection of measurements.

mantissa The decimal part of a logarithm (Appendix 3) or of a number containing an exponent of ten, e.g. in 3.4×10^5 the mantissa is 3.4.

matrix A table of quantities (usually numbers) subject to certain mathematical rules.

mean The average. The sum of a collection of numerical measurements divided by the number of measurements.

median That value of the variate which divides the total frequency into two equal halves.

mode The value of the variate possessed by the greatest number of the population. Often used as a measure of location to denote the class interval containing the largest number of observations.

modulus The absolute value of a number, given a positive sign and usually placed between vertical bars: | |.

nominal A variable which cannot be measured and must be expressed qualitatively. Also known as attribute, attribute variable or categorical variable. The nominal scale is the weakest level of measurement.

non-parametric A distribution-free statistical test which can be used to analyse a wider range of data than parametric tests (see parameter).

numerator In a fraction the value which is divided by the denominator, e.g. the number 7 in 7/9.

ordinate The y coordinate or y axis measured up the page of a graph from the origin.

ordination The reduction in dimensionality of multivariate data, usually displayed as a scattergraph showing the relationships between the samples or variables.

origin The point on a graph where the x and y axes meet and where $x = 0$, $y = 0$.

outliers Values which lie well outside the range of the remaining measurements of a sample. They are often the result of incorrect measurements. Also called mavericks and wildshots.

parameter In statistics it is a value which helps define a frequency distribution of a population. Parameters are usually fixed for a particular distribution model, e.g. the population parameters μ and σ of the normal distribution. Parametric statistical tests require the estimation of the parameters and are subject to a range of assumptions concerning the distribution of the data.

population A collection of items from which samples are drawn. The size of a population may be finite or infinite. Usually, its exact size is unknown and it may or may not be represented by a normal distribution.

precision The closeness of repeated measurements of the same quantity.

quartile The three values, upper and lower quartile and median, which divide the frequency of a distribution into four equal parts.

range The largest minus the smallest of a set of sample values. An elementary measure of dispersion.

rank The ordinal number in a sample where the values of the variate are ordered from lowest to highest (or, occasionally, highest to lowest).

ratio A scale of measurement amenable to arithmetic procedures and with an absolute zero point. Also used for the quotient of two numbers (as in analysis of variance).

regression line Line obtained statistically, relating two (or more) variables to each other. The line is usually straight, but may also be curved.

residual The difference between an observed number (y_i) and a predicted value ($\hat{y}$) obtained from a statistical model.

sample A part of a population, often selected to investigate properties of the parent population or set.

skew Indicating a lack of symmetry in curves and frequency histograms.

standard deviation A measure of dispersion equal to the square root of the variance.

subscript A number shown below a variable, e.g. x_3, denoting its position in an ordering scheme.

sum of squared deviations Usually abbreviated to *sum of squares* and denoting the sum of (deviations from the mean)^2.

superscript A number shown above a variable, e.g. x^2, and normally denoting the power to which the variable is raised (exponent).

synergism Used when the combined effect of two different treatments is greater than the sum of the effects of the individual treatments.

tie Term applied to ranks where two or more ranks are assigned the same value.

variable An entity which can stand for any of the members of a given set. The numbers of the set constitute values and the set defines the variables' range. A variate differs in that the variable is always associated with a probability distribution.

variance The sum of squared deviations divided by the number of observations (n) is a sample variance, but it is a biased estimator of the population variance. The sum of squares divided by $n - 1$ is an unbiased estimator of the population variance and is the quantity usually calculated as the variance in statistical analysis.

variate A random variable. A variable that can take any one of a range of values each with an associated probability.

Index